NATIONAL ACADEMIES *Sciences Engineering Medicine*

NATIONAL ACADEMIES PRESS
Washington, DC

Fostering Responsible Computing Research

Foundations and Practices

Committee on Responsible Computing Research: Ethics and Governance of Computing Research and Its Applications

Computer Science and Telecommunications Board

Division on Engineering and Physical Sciences

Consensus Study Report

THE NATIONAL ACADEMIES PRESS 500 Fifth Street, NW Washington, DC 20001

This activity was supported by the National Science Foundation Grant No. CNS-1937181. Any opinions, findings, conclusions, or recommendations expressed in this publication do not necessarily reflect the views of any organization or agency that provided support for the project.

International Standard Book Number-13: 978-0-309-29527-7
International Standard Book Number-10: 0-309-29527-0
Digital Object Identifier: https://doi.org/10.17226/26507

This publication is available from the National Academies Press, 500 Fifth Street, NW, Keck 360, Washington, DC 20001; (800) 624-6242 or (202) 334-3313; http://www.nap.edu.

Copyright 2022 by the National Academy of Sciences. National Academies of Sciences, Engineering, and Medicine and National Academies Press and the graphical logos for each are all trademarks of the National Academy of Sciences. All rights reserved.

Printed in the United States of America.

Suggested citation: National Academies of Sciences, Engineering, and Medicine. 2022. *Fostering Responsible Computing Research: Foundations and Practices*. Washington, DC: The National Academies Press. https://doi.org/10.17226/26507.

The **National Academy of Sciences** was established in 1863 by an Act of Congress, signed by President Lincoln, as a private, nongovernmental institution to advise the nation on issues related to science and technology. Members are elected by their peers for outstanding contributions to research. Dr. Marcia McNutt is president.

The **National Academy of Engineering** was established in 1964 under the charter of the National Academy of Sciences to bring the practices of engineering to advising the nation. Members are elected by their peers for extraordinary contributions to engineering. Dr. John L. Anderson is president.

The **National Academy of Medicine** (formerly the Institute of Medicine) was established in 1970 under the charter of the National Academy of Sciences to advise the nation on medical and health issues. Members are elected by their peers for distinguished contributions to medicine and health. Dr. Victor J. Dzau is president.

The three Academies work together as the **National Academies of Sciences, Engineering, and Medicine** to provide independent, objective analysis and advice to the nation and conduct other activities to solve complex problems and inform public policy decisions. The National Academies also encourage education and research, recognize outstanding contributions to knowledge, and increase public understanding in matters of science, engineering, and medicine.

Learn more about the National Academies of Sciences, Engineering, and Medicine at **www.nationalacademies.org**.

Consensus Study Reports published by the National Academies of Sciences, Engineering, and Medicine document the evidence-based consensus on the study's statement of task by an authoring committee of experts. Reports typically include findings, conclusions, and recommendations based on information gathered by the committee and the committee's deliberations. Each report has been subjected to a rigorous and independent peer-review process and it represents the position of the National Academies on the statement of task.

Proceedings published by the National Academies of Sciences, Engineering, and Medicine chronicle the presentations and discussions at a workshop, symposium, or other event convened by the National Academies. The statements and opinions contained in proceedings are those of the participants and are not endorsed by other participants, the planning committee, or the National Academies.

Rapid Expert Consultations published by the National Academies of Sciences, Engineering, and Medicine are authored by subject-matter experts on narrowly focused topics that can be supported by a body of evidence. The discussions contained in rapid expert consultations are considered those of the authors and do not contain policy recommendations. Rapid expert consultations are reviewed by the institution before release.

For information about other products and activities of the National Academies, please visit www.nationalacademies.org/about/whatwedo.

COMMITTEE ON RESPONSIBLE COMPUTING RESEARCH: ETHICS AND GOVERNANCE OF COMPUTING RESEARCH AND ITS APPLICATIONS

BARBARA J. GROSZ, NAE,[1] Harvard University, *Chair*
MARK ACKERMAN, University of Michigan, Ann Arbor
STEVEN M. BELLOVIN, NAE, Columbia University
MARIANO-FLORENTINO CUÉLLAR, Carnegie Endowment for International Peace
DAVID DANKS, University of California, San Diego
MEGAN FINN, University of Washington
MARY L. GRAY, Microsoft Research
JOHN L. HENNESSY, NAS[2]/NAE, Stanford University and Alphabet, Inc.
AYANNA M. HOWARD, The Ohio State University
JON M. KLEINBERG, NAS/NAE, Cornell University
SETH LAZAR, Australian National University
JAMES MANYIKA, McKinsey Global Institute and Google, Inc.
JAMES MICKENS, Harvard University
AMANDA STENT, Colby College

Staff

JON K. EISENBERG, Senior Board Director, Study Director
KATIRIA ORTIZ, Associate Program Officer
SHENAE A. BRADLEY, Administrative Assistant

---

[1] Member, National Academy of Engineering.
[2] Member, National Academy of Sciences.

COMPUTER SCIENCE AND TELECOMMUNICATIONS BOARD

LAURA HAAS, NAE,[1] University of Massachusetts Amherst, *Chair*
DAVID CULLER, NAE, University of California, Berkeley
ERIC HORVITZ, NAE, Microsoft Research
CHARLES ISBELL, Georgia Institute of Technology
ELIZABETH MYNATT, Georgia Institute of Technology
CRAIG PARTRIDGE, Colorado State University
DANIELA RUS, NAE, Massachusetts Institute of Technology
MARGO SELTZER, NAE, University of British Columbia
NAMBIRAJAN SESHADRI, NAE, University of California, San Diego
MOSHE Y. VARDI, NAS[2]/NAE, Rice University

Staff

JON K. EISENBERG, Senior Board Director
SHENAE A. BRADLEY, Administrative Assistant
RENEE HAWKINS, Finance Business Partner
THƠ NGUYỄN, Senior Program Officer
KATIRIA ORTIZ, Associate Program Officer
BRENDAN ROACH, Program Officer

---

[1] Member, National Academy of Engineering.
[2] Member, National Academy of Sciences.

Preface

Computing technology is increasingly woven into our personal and professional lives, physical infrastructure, and societal fabric. With this rise in computing's impact comes an interest in ensuring that its use contributes to human flourishing; thriving societies; a healthy planet; and an interest in addressing ethical and societal impact concerns that arise when computing technologies lead to such undesirable outcomes as an erosion of personal privacy, the spread of false information and propaganda, biased or unfair decision-making, disparate socioeconomic impacts, or diminished human agency.

It has become increasingly apparent that it is vital for the computing research community to increase its capacity to address these concerns. Accordingly, the National Science Foundation requested that the National Academies of Sciences, Engineering, and Medicine examine best practices that research sponsors, research-performing institutions, and individual researchers can use to formulate, conduct, and evaluate computing research and associated activities in a responsible manner (see Box P.1).

To carry out the study, the National Academies appointed the Committee on Responsible Computing Research (see Appendix A). The study committee comprised expertise across many areas of computer science and engineering, information science, computing technology development, social sciences, philosophy, and law. Within computer science and engineering, the committee included expertise in an array of subfields: theory, systems, artificial intelligence, human–computer interaction, cybersecurity, and robotics. Based on the Statement of Task, the committee has focused on practical approaches based on scholarship in ethics and in scholarship on sociotechnical systems together with approaches from computer science and engineering, information science, and related fields such as design.

> **BOX P.1 Statement of Task**
>
> A National Academies study will explore ethics and governance issues associated with the personal and social consequences of computing research and its applications. The study committee will gather input through at least one open meeting and a solicitation for written comments from relevant research communities and stakeholders. It will consider such topics as:
>
> 1. Guiding principles, tools, and practical approaches for identifying and addressing ethical issues;
> 2. The feasibility and likely performance of research governance frameworks and regulatory regimes, and related best practices that research funders, research-performing institutions, and individual researchers can leverage to formulate, conduct, and evaluate ethical research and associated activities;
> 3. Multidisciplinary approaches to understanding ethical issues in computing research;
> 4. How these approaches can empower the research community to develop and pursue socially productive practices; and
> 5. Ways to promulgate ethical principles and responsible practices and sustain attention to them in the computing research community, including through education and training.
>
> The study will consider these issues across different subdomains or application areas of computing, such as medicine, autonomous vehicles, and elections. The study will not focus on ethical issues associated with the conduct of research itself except where these relate to the implications of research results.
>
> In carrying out this study, the committee will also consider related questions such as: (a) How do ethics and governance issues and needs present differently in different research contexts? Are there other ethics and governance issues that apply more broadly across many or most areas of computing research? (b) What set of research governance frameworks or regulatory regimes are feasible in each of these contexts? (c) How might research governance take place at different granularities and modalities of governance, such as community, organizational, local, regional, national, and international? (d) What empirical evidence exists for how these research governance frameworks or regulatory regimes might correspond to ethically desirable outcomes? (e) What is the current relative maturity level of ethics and governance concepts in different aspects of the computing research space? Which areas are the most advanced and can their relative maturity be leveraged into use elsewhere in computing? (f) What incentives or contextual changes would be effective in helping computing researchers, and those who develop subsequent applications, place more emphasis on ethical considerations? For which existing, and likely future, stakeholders are such changes compatible with current incentives?
>
> The committee will prepare a final report containing its analysis, findings, and (as appropriate) recommendations. The report will identify and (to the extent feasible) recommend practical steps that National Science Foundation–supported researchers and others in the computing research community can take to address ethics in all phases of their research from proposal to publication.

Several members of the committee changed their primary professional affiliations during the course of this study. Mariano-Florentino Cuéllar, previously justice of the Supreme Court of California and Herman Phleger Visiting Professor at Stanford University, became president of the Carnegie Endowment for International Peace; David Danks, previously L.L. Thurstone Professor of Philosophy and Psychology at Carnegie Mellon University, became professor of data science and philosophy at the University of California, San Diego; James Manyika, previously chairman and director of the McKinsey Global Institute, became senior vice president of technology and society at Google, Inc.; and Amanda Stent, previously NLP Architect at Bloomberg LP, became director of the Davis Institute for Artificial Intelligence at Colby College. Also, Alondra Nelson, Harold F. Linder Professor at the Institute for Advanced Study, stepped down from the committee in January 2021 when she was appointed as deputy director for science and society at the White House Office of Science and Technology Policy.

In order to explore ethical and societal impact issues in context, the committee convened public meetings with experts in criminal and civil justice, public governance, work and labor, and health care and with research managers from the computing industry and federal sponsors of computing research (see Appendix B); the committee benefited greatly from the insights these experts contributed. The committee did not consider the distinctive trade-offs associated with the context of national security, cognizant that other groups with more focused expertise have extensively examined such matters.[1]

Early on in its deliberations, the committee considered the questions in the Statement of Task about governance and regulatory regimes and quickly realized that few if any of these are in place today. What does exist today are sets of principles aimed at guiding those engaged in developing and deploying computing technologies. These principles may be a useful starting point but, as is discussed in Chapter 2 of this report, they are insufficient in themselves as they are divorced from practice and do not provide sufficiently thorough explanations of their underlying assumptions.

The primary aim of this report and its recommendations is to empower the computing research community to further develop and use these practical approaches and attain socially beneficial research practices. The committee believes that the adoption of such practices at the research stage will have significant downstream effects by serving as a model for those who develop and deploy computing technologies. Toward this end,

---

[1] For example, National Security Commission on Artificial Intelligence, 2021, *Final Report*, https://www.nscai.gov/2021-final-report, and National Research Council and National Academy of Engineering, 2014, *Emerging and Readily Available Technologies and National Security: A Framework for Addressing Ethical, Legal, and Societal Issues*, The National Academies Press, Washington, DC.

the report considers needs for reshaping not only research practice but also computing education; the recommended changes in computing education will help ensure that future computing professionals across the industry are better equipped to address ethical and societal concerns. Last, the recommendations also include measures that could help reshape incentives in the computing research ecosystem so that they are better aligned with the goal of responsible computing research.

Acknowledgment of Reviewers

This Consensus Study Report was reviewed in draft form by individuals chosen for their diverse perspectives and technical expertise. The purpose of this independent review is to provide candid and critical comments that will assist the National Academies of Sciences, Engineering, and Medicine in making each published report as sound as possible and to ensure that it meets the institutional standards for quality, objectivity, evidence, and responsiveness to the study charge. The review comments and draft manuscript remain confidential to protect the integrity of the deliberative process.

We thank the following individuals for their review of this report:

Elizabeth Bradley, University of Colorado Boulder,
Kenneth Calvert, University of Kentucky,
Deborah Crawford, University of Tennessee, Knoxville,
Finale Doshi-Velez, Harvard University,
Batya Friedman, University of Washington,
Eric Horvitz, NAE,[1] Microsoft,
Charles Isbell, Georgia Institute of Technology,
Craig Partridge, Colorado State University,
Fernando Pereira, NAE, Google, Inc.,

---

[1] Member, National Academy of Engineering.

Allison Stanger, Middlebury College, and
Moshe Vardi, NAS[2]/NAE, Rice University.

Although the reviewers listed above provided many constructive comments and suggestions, they were not asked to endorse the conclusions or recommendations of this report nor did they see the final draft before its release. The review of this report was overseen by the monitor, Samuel H. Fuller, NAE, Analog Devices, Inc. He was responsible for making certain that an independent examination of this report was carried out in accordance with the standards of the National Academies and that all review comments were carefully considered. Responsibility for the final content rests entirely with the authoring committee and the National Academies.

---

[2] Member, National Academy of Sciences.

Contents

SUMMARY   1

1   INTRODUCTION   9
     1.1 The Nature of Computing and Computing Research, 9
     1.2 The Nature of the Technology Innovation Ecosystem, 11
     1.3 The Nature of the Computing Research Ecosystem, 12
     1.4 The Roles of Ethics and Social Science in Computing, 14
     1.5 Sources of Ethical and Societal Impact Challenges, 15
     1.6 A Brief History of Concerns, 16
     1.7 Characteristics of Responsible Computing in Light of the Ubiquity of Computing Technologies, 19
     1.8 Study Approach, 20

2   THEORETICAL FOUNDATIONS FROM ETHICAL AND SOCIAL SCIENCE FRAMEWORKS   23
     2.1 The Value and Scope of Ethics, 24
     2.2 The Power of a Sociotechnical Perspective, 33

3   SOURCES OF ETHICAL CHALLENGES AND SOCIETAL CONCERNS FOR COMPUTING RESEARCH   44
     3.1 Societal Contexts, 46
     3.2 Limitations of Human Capabilities, 61
     3.3 Societal Contexts and Design and Deployment Choices, 66
     3.4 System Robustness, 87
     3.5 Limits of a Purely Computing-Technical Approach, 97

4   CONCLUSIONS AND RECOMMENDATIONS   103
     4.1 Reshape Computing Research, 108
     4.2 Foster and Facilitate Responsible Computing Research, 110
     4.3 Support the Development of the Expertise Needed to Integrate Social and Behavioral Science and Ethical Thinking into Computing Research, 113
     4.4 Ensure That Researchers Have Access to the Knowledge and Expertise Needed to Assess the Ethical and Societal Implications of Their Work, 117

  4.5 Integrate Ethical and Societal Considerations into Computing Research Sponsorship, 119

  4.6 Integrate Ethical and Societal Considerations into Publication, 123

  4.7 Adhere to Best Practices for Systems Design, Deployment, Oversight, and Monitoring, 127

  4.8 Support Engagement with the Public and the Public Interest, 129

APPENDIXES

A	**Committee Member Biographical Information**	135
B	**Presentations to the Committee**	142
C	**Federal Computing Research Programs Related to Ethical and Societal Impact Concerns**	145

Summary

Computing technologies were once the purview of computing researchers and professionals, with their use largely confined to corporate, defense, laboratory, or other closed environments with a limited number of users. Today, they are woven into our personal and work lives; our economic, social, educational, and political institutions; and the constructed environment around us. Computing research ideas are continually integrated into existing systems and regularly lead to entirely new technologies. The impacts of new developments in computing technology are often hard to predict and larger than anticipated. Few fields rival computing for the speed with which research advances are deployed and used by millions of people.

Computing research—a term used in this report to include research in computer science and engineering, information science, and related fields—thus cannot simply focus narrowly on development of innovative computing methods and systems. Responsible research requires that the ethical and societal impacts of computing research and the technologies the research enables be first-order concerns of the computing research community.

To ensure that computing research addresses these impacts and concerns and supports human flourishing, thriving societies, and a healthy planet, computing researchers must consult and collaborate with scholars and researchers in the humanities and social and behavioral sciences, particularly but not only those who study moral reasoning or the empirical analysis of sociotechnical systems or who can help cultivate moral imagination about alternative outcomes. An understanding of fundamental ethical concepts (see Section 2.1) will enable computing researchers to engage in careful analysis and decision-making about ethical challenges. Likewise, an understanding of the sociotechnical approach (see Section 2.2)—which draws on social theories and social scientific research methodologies—will enable computing researchers to identify the benefits and risks that accompany the introduction of novel technologies into societal contexts. Importantly, this does not mean that computer scientists, computer engineers, and information scientists are expected to become expert ethicists and social scientists but rather that they should collaborate with experts in other disciplines who can bring this expertise to bear as computing research is designed and carried out.

Failure to consider consequences early in research increases the risk of adverse ethical or societal impacts. Scholarship in the field of design (see Section 3.4.4) has developed approaches that enable principled considerations of potential consequences and envisioning alternatives in the design space. A well-known design principle in computing illuminates the importance of considering ethical and societal impact issues in research: it is much easier to design a technology correctly from the start than it is to fix it later.

To identify and explore potential sources of ethical challenges and societal concerns associated with computing research, the study committee heard from experts at the intersection of computing and the domains of health care, work and labor, civil and criminal justice, and governance. The committee also heard from industry and government research managers. Combining insights gleaned from these experts and the theoretical foundations from ethics and sociotechnical systems yielded a set of illustrative ethical and societal concerns for computing research that are discussed in Chapter 3. These examples of concerns fall into four groups; those that arise from (1) features of the societal settings into which new computing technologies are introduced; (2) limitations of human capabilities and ways they interact with features of computing technologies and the societal contexts in which computing technologies may be used; (3) features of the societal contexts into which computing technologies may be integrated that influence computing system design and deployment choices; and (4) system robustness problems caused by failure to follow best practices in design and implementation. For each concern, there are opportunities and obligations for computing researchers in collaboration with social and behavioral scientists and ethicists to address them. Because many of the problems uncovered arise from misinterpretation or unintended uses of research results, it is incumbent on researchers to act to minimize the possibility of misinterpretation or misuse of their research results.

These concerns point to three conclusions that underlie the recommendations that follow:

Conclusion 1. To be responsible, computing research needs to expand to include consideration of ethical and societal impact concerns and determination of effective ways to address them.

Conclusion 2. To be responsible, computing research needs to engage the full spectrum of stakeholders and deploy rigorous methodologies and frameworks that have proven effective for identifying the complicated social dynamics that are relevant to these ethical and societal impact concerns.

Conclusion 3. For computing technologies to be used responsibly, governments need to establish policies and regulations to protect against adverse ethical and societal impacts. Computing researchers can assist by revealing limitations of their research results and identifying possible adverse impacts and needs for government intervention.

This third conclusion stems from the observation that the design and deployment of computing technologies are shaped by a combination of corporate decision-making, incentives set by the market and government regulation, and decisions made by organizations in acquiring the technologies. These factors are the proper realm of societies, which determine norms, and of governments, which institute mechanisms to enforce those norms. Nevertheless, computing researchers have responsibilities related to societal and ethical concerns arising from these technologies.

The report's recommendations are collectively aimed at all actors in the computing research ecosystem, including researchers; organizations that sponsor and carry out research in academia, industry, and government; scientific societies; and scholarly publishers. They define practices to help foster responsible computing research, including identification and mitigation of potential harms as well as promotion of research providing ethical and societal benefits.

There are eight high-level recommendations, listed below. Each recommendation is accompanied by supporting subrecommendations indicating practical steps to be taken to implement the recommendation. These subrecommendations are summarized below and listed in full and discussed further in Chapter 4.

Several of the recommendations necessitate adapting processes and approaches in the research ecosystem or developing new ones to meet the needs of responsible computing research. There is, as yet, little if any empirical data on the performance of different practical steps for incorporating considerations of ethical and societal impacts in computing research or approaches to responsible computing research more generally. Acquiring such data is necessary for advancing responsible computing research throughout the ecosystem. As with innovation in science and engineering, the innovations called for in these recommendations therefore require ongoing assessment and revision to determine what works best. Subrecommendations to do so are thus included under the relevant top-level recommendations.

> **Recommendation 1. The computing research community should reshape the ways computing research is formulated and undertaken to ensure that ethical and societal consequences are considered and addressed appropriately from the start.**

In developing and carrying out their projects, researchers should include participants with germane expertise in the social and behavioral sciences, ethics, and any domains of application involved; and if their projects aim for societally relevant outcomes, they should involve relevant stakeholders throughout the research. Publications and other accounts of their research should discuss possible limitations and any downstream risks of artifacts as well as algorithms and other computing methods.

> **Recommendation 2. The computing research community should initiate projects that foster responsible computing research, including research that leads to societal benefits and ethical societal impact and research that helps avoid or mitigate negative outcomes and harms. Both research sponsors and research institutions should encourage and support the pursuit of such projects.**

To advance such responsible, societally beneficial research projects and facilitate the multidisciplinary research called for in Recommendation 1, research sponsors and universities should support new kinds of projects and new types of partnerships with companies and philanthropies. To enable and incentivize researchers to participate in such research, (1) research sponsors should provide sufficient resources for the participation of scholars from fields other than computer science and engineering and of stakeholders, and (2) research institutions' performance review processes and tenure and promotion committees should value both disciplinary and multidisciplinary scholarship on the ethical and societal impacts of computing research.

> **Recommendation 3. Universities, scientific and professional societies, and research and education sponsors should support the development of the expertise needed to integrate social and behavioral science and ethical thinking into computing research.**

Academic institutions should reshape their computer science and engineering curricula and curricula in social and behavioral sciences and the humanities to better equip students to address the ethical and societal impacts of computing, and to support their graduates' abilities to assist public- and private-sector institutions in making better decisions as they acquire computing technologies. Concomitantly, scientific and professional societies as well as research sponsors should provide training opportunities for computing researchers to gain proficiency in carrying out and assessing responsible computing research.

Recommendation 4. Computing research organizations—working with scientific and professional societies and research sponsors—should ensure that their computing faculty, students, and research staff have access to scholars with the expertise to advise them in examining potential ethical and societal implications of proposed and ongoing research activities, including ways to engage relevant groups of stakeholders. Computing researchers should seek out such advice.

To buttress responsible computing research efforts, research institutions, scientific societies, and research sponsors should develop ways for computing researchers to identify scholars with the ethical, societal impact, and domain expertise their projects require and provide support for such scholars to collaborate in the research. Research sponsors also should support the development and sharing of educational materials and descriptions of best practices.

Recommendation 5. Sponsors of computing research should require that ethical and societal considerations be interwoven into research proposals, evaluated in proposal review, and included in project reports.

Research proposals should describe in an integrated fashion the ethical and societal considerations associated with the proposed work and not in a separate section. Research sponsors should ensure that project review panels are provided with appropriate evaluation criteria and have the requisite expertise to evaluate these considerations, and they should require that project reports address ethical and societal issues that arise.

Recommendation 6. Scientific and professional societies and other publishers of computing research should take steps to ensure that ethical and societal considerations are appropriately addressed in

publications. The computing research community should likewise take steps to ensure that these considerations are appropriately addressed in the public release of artifacts.

Conferences and journals should establish evaluation criteria and metrics to be used in assessing a paper's treatment of ethical and societal impacts and provide guidelines for authors and reviewers. They should encourage researchers to report unanticipated ethical or societal consequences of their research and to provide guidance to future researchers interested in using the results of their research. They should also establish criteria for determining whether and how to release artifacts.

Recommendation 7. Computing researchers who are involved in the development or deployment of systems should adhere to established best practices in the computing community for system design, oversight, and monitoring.

Researchers should also be transparent about the capabilities, maturity, and limitations of any artifacts they produce and document their design assumptions.

Recommendation 8. Research sponsors, research institutions, and scientific and professional societies should encourage computing researchers to engage with the public and with the public interest and support them in doing so.

Specific opportunities for such engagement include informing the public, assisting public- and private-sector acquirers of computing technologies, and bringing potential adverse consequences of emerging technologies to the attention of governments and other public organizations. In addition, universities, research sponsors, and scientific societies should create opportunities for computing researchers to learn how to effectively serve in advisory capacities.

* * *

Computing researchers will not be able to eliminate every potential ethical or societal problem in the computing research ecosystem. They can, however, be proactive in contributing to the benefit of society and in identifying risks and avoiding potential harms. Doing so requires that they broaden the scope of computing research in ways that these recommendations propose. It also requires that assessments of computing research encompass not only performance analysis and mathematical advances but also evaluation of potential ethical issues and societal impacts—thus requiring research organizations and scientific and professional societies to make changes as well. The recommendations are also intended to have downstream impacts. First, researchers following the recommendations will provide a model not only for other researchers but also for technology developers and deployers. Second, the recommendations for changes in computing education will help ensure that future computing professionals across industry, not just in research, are better equipped to address ethical and societal concerns.

1 Introduction

1.1 THE NATURE OF COMPUTING AND COMPUTING RESEARCH

The fruits of computing research—a term used in this report to include research in computer science and engineering, information science, and related fields—have been increasingly woven into our personal and work lives and the constructed world around us, making us all stakeholders in how computing is used. The innovations enabled by computing research have improved the lives of individuals, the work of institutions and organizations of various sorts, and the functioning of governments and communities. For example, automobiles have been made safer with the introduction of antilock brakes and pedestrian detection, better weather forecasts help save lives and enhance crop yields, and medical discovery has been accelerated by ever-greater raw computational power along with advances in data analytics and modeling.

At the same time, as regularly reported in both the general and scientific press, some uses of computing technology have raised concerns about individual and societal harms. For instance, the use of predictive analytics in administering criminal justice risks perpetuating structural biases in society. New communication and entertainment platforms have afforded new avenues for spreading misinformation and disinformation. Concerns such as these and others have led to increasing calls for governments to act both in making wiser decisions about their own use of computing technology and in revising regulations or implementing new ones. Furthermore, computing research has an obligation to support human flourishing, thriving societies, and a healthy planet.

Unlike research in the natural sciences, computing research creates and studies computing artifacts—computer hardware and software and associated data, models, and algorithms—that are all human-made things. Computing research thus continually creates new possibilities for human action and is fundamentally a human-inspired (and largely human-constrained) endeavor.

Although computing research encompasses engineering, it differs from other engineering endeavors because it is limited more by human imagination than by the physical constraints often found in other areas of engineering. Frederic Brooks made similar observations about the nature of software programming in the title essay of *The Mythical Man-Month*:

> In many creative activities the medium of execution is intractable. Lumber splits; paints smear; electrical circuits ring. These physical limitations of the medium constrain the ideas that may be expressed, and they also create unexpected difficulties in the implementation.... Computer programming, however, creates with an exceedingly tractable medium. The programmer builds from pure thought-stuff: concepts and very flexible representations thereof.[1]

There are, of course, some fundamental physical limits, such as energy, heat dissipation, and integrated circuit feature size, and some limits on what can theoretically be computed in a reasonable amount of time. Moreover, there are important resource and environmental constraints on what one *should* build—see Section 3.1.6.

To be sure, the nature of the medium is not quite as unconstrained as the quote above suggests. Not everything can be quantified or represented formally, and a particular programming language, like any technology, makes it easier to do some things and harder to do other things. Nonetheless, algorithms are extremely flexible, and computing technologies are used in many disciplines, scientific and otherwise, and every sector of society. Many algorithms, models, and pieces of software can be used for a wide variety of purposes, some beyond their original design intentions. The barrier to embedding software and hardware in the real world is low and continually decreasing. Anyone with basic programming skills can build and deploy an app. This general-purpose nature means that computing research can affect a wide diversity of applications, contexts, and societal domains. Computing researchers thus need to be especially thoughtful, creative, and diligent when considering potential societal and ethical implications of their work, and they need to be assiduous in describing the intended uses and limitations of that work.[2]

---

[1] F.P. Brooks, Jr., 1975, *The Mythical Man-Month*, Anniversary Edition, Addison-Wesley Longman, Boston, MA, 1995.

[2] See, for example, F. Urbina, F. Lentzos, C. Invernizzi, and S. Ekins, 2022, "Dual Use of Artificial-Intelligence-Powered Drug Discovery," *Nature Machine Intelligence* 4:189-191, https://doi.org/10.1038/s42256-022-00465-9.

1.2 THE NATURE OF THE TECHNOLOGY INNOVATION ECOSYSTEM

To understand what is needed for responsible computing research, one needs to understand the vibrant technology ecosystem in which computing research takes place. This ecosystem comprises universities (with a majority of support from government research sponsors), both small and large computing firms, and government laboratories. Especially in the United States, the ecosystem features rich interplay among academic researchers, industry researchers, and the creators of products and services. This ecosystem benefits greatly from multi-directional flows of ideas, artifacts, technologies, and people. These multidirectional flows speed up the pace of deployment of computing research, fueling transformative results across every sector of the U.S. economy.[3]

The time required for basic research ideas to have commercial impact varies widely. Sometimes it takes years or even decades, as innovations compound and mature, technology components become less expensive, and market needs emerge. Sustained and patient research investment is often needed to realize the full potential of computing research. For example, machine learning research started more than five decades ago, saw its first significant commercial applications in the early 1990s, and dramatically accelerated in impact a decade ago enabled by a combination of new algorithms, new sources of training data from an increasingly interconnected and digitized society, and advances in computing hardware.[4]

Other times, the rich connections in the technology innovation ecosystem and the availability of funding from venture capital or well-resourced firms make it possible for research ideas to be deployed quickly and on a large scale.[5] One recent example is the use of machine learning to predict advertising clickthrough rates. The idea made its way from a published paper to deployment by Facebook in 6 months.[6] Few if any fields rival computing for the speed with which research advances can be deployed to be used by millions of people.

That being said, the transfer of an idea from a research lab to a consumer-facing product is a dynamic process that is often beset by unexpected challenges. Predeployment, algorithms from a paper may be tweaked, extended, or eliminated from a system

---

[3] NASEM (National Academies of Sciences, Engineering, and Medicine), 2020, *Information Technology Innovation: Resurgence, Confluence, and Continuing Impact*, The National Academies Press, Washington, DC, https://doi.org/10.17226/25961.

[4] Ibid.

[5] Ibid.

[6] The original paper is T. Graepel, J. Quiñonero Candela, T. Borchert, and R. Berbrich, 2010, "Web-Scale Bayesian Click-Through Rate Predication for Sponsored Search Advertising in Microsoft's Bing Search Engine," Pp. 13-20 in *Proceedings of the 27th International Conference on Machine Learning*. https://www.microsoft.com/en-us/research/wp-content/uploads/2010/06/AdPredictor-ICML-2010-final.pdf. For a history of how this work was implemented at Facebook, see K. Hao, 2021, "How Facebook Got Addicted to Spreading Misinformation," https://www.technologyreview.com/2021/03/11/1020600/facebook-responsible-ai-misinformation.

to better fit the exigencies of real-world environments. Post-deployment, systems may be altered in response to relevant advances from the research community, or because unexpected dangers are uncovered. The net result is that the technology transfer of an idea from research to industry is not a single-shot event, but rather a feedback process that unfolds over time and has many inputs. This "continuous integration" means that the work of computing researchers can have significant impacts on deployed systems even long after the systems are launched, and that research continues to be linked to downstream outcomes.

The computing industry has considerable experience in translating research into practice. Companies have built processes and developed in-house expertise for identifying potential issues with new technologies that they can bring to bear as new technologies are deployed or integrated into products. These experiences and practices can be instructive to the computing research community.

Testing of the systems produced by computing research occurs partly under controlled conditions and partly in uncontrolled environments with unknown users. The latter has become increasingly consequential as computing technology is deployed more widely (at a larger scale and to more diverse populations). Full testing of an approach or artifact in the context of a particular application inherently involves people using the system in that situation (e.g., nurses using a new electronic health records system in a hospital). Doing this testing can be costly (the difficulty of recruiting the "human subjects") and complicated (because the testing itself may raise ethical issues). These issues create challenges for responsible computing research.

The ways a new computing technology is ultimately used may also differ greatly from the original intent of its inventors. For example, Web cookies were introduced in the mid-1990s as a way of maintaining state without storing information on a server so that users did not have to keep reentering the same information. Within a short time, however, third-party cookies were introduced as a way of tracking user activity across websites, almost immediately raising privacy concerns about a technology that was originally thought to be privacy protecting. A related issue is that algorithms or actual software artifacts are frequently used in application areas other than those contemplated in the original research.

1.3 THE NATURE OF THE COMPUTING RESEARCH ECOSYSTEM

Many actors participate in the multi-step translation of research results into deployed algorithms, devices and systems, including researchers, research sponsors, entrepreneurs, investors, and corporate leaders. The roles played by the various participants in the research enterprise vary as do the range of interactions they have with others, their capabilities for influencing outcomes, and the incentive structures that influence their choices. Indeed, incentive structures play a large role throughout the computing research ecosystem, from the Ph.D. student incentives for graduating and obtaining good positions, to faculty concerned about tenure and research funding, to startups aiming to establish a beachhead in the market, to giant corporations' incentives for maintaining their market position.

A well-known design principle in computing illuminates the importance of considering ethical and societal impact issues in research: it is much easier to design a technology correctly from the start than it is to fix it later. Furthermore, choices among research topics and research methods are determinative of possible computing technologies. In focusing on computing research, this study considers how ethical and societal impact challenges can be addressed at this consequential foundational stage, and the practical steps that computing researchers, the research community as a whole, research sponsors, and research-performing institutions can take toward fostering the development of computing technologies that more often serve social good and less often cause harm. Scientists and engineers have another important role as well: informing and educating future computing professionals about ethical and societal impact responsibilities and ways to meet them. The changes in computing education will make it more likely that the computing industry deploys computing technologies in an ethical and responsible manner.

Still, computing researchers cannot, by themselves, ensure that computing technologies are designed and used in ways that are ethical and responsible. Technologies can have many different uses, and it is the application and context of use that most directly create the ethical and societal impact challenges. Moreover, most innovations that lead to new capabilities, including those raising ethical and societal challenges, draw on a combination of many research results.

Thus, science and engineering are necessary, but they alone do not suffice to address ethical issues, because, as Chapter 2 discusses, the design of computing technologies involves many trade-offs among values, preferences, and incentive choices. Such public policy choices (which include decisions about what technologies government acquire) are the proper realm of societies and communities, not of scientists and engineers, in determining norms and governments in instating mechanisms to realize or enforce those norms. Nonetheless, scientists and engineers have important roles to play in ensuring that government decision-makers have the information they need about the results of research to make wise choices. The recommendations of this study also aim to inform, complement, and support actions by government and companies.

How should computer scientists think about these challenges? This report argues that computer scientists must begin to treat ethical and societal impact considerations as first-order concerns. In the same way that computer scientists understand that quantitative metrics such as classification accuracy, algorithm efficiency, and energy consumption are important, computer scientists must now also reckon with the ethical and societal impacts of deployed technologies that incorporate their research results. The report recommends practical steps toward doing so.

1.4 THE ROLES OF ETHICS AND SOCIAL SCIENCE IN COMPUTING

Ethics provides tools for the moral evaluation of behaviors, institutions, and social structures and for dealing with choices among and conflicts between values. Until relatively recently, many researchers and observers considered computing technologies to be value neutral. Few if any are. The design of new computing technologies, much as with technologies more generally, is always imprinted with the spectrum of values considered by the designer, which may not be broad enough to ensure a particular technology meets the needs of some stakeholders. Sometimes the value choices are intended. Very often they are not.[7] Some of these values may be explicitly expressed while others may be implicit. Moreover, some of these values may be introduced during research, not solely in the translation of research into viable technologies. All technologies, and the research that enables them, create some opportunities and foreclose other possibilities.

Ethics is concerned with doing good as well as avoiding harm. Consideration of ethical and societal impacts in computing research thus includes both proactive research to create computing technologies that do (more) good and preventative work to anticipate, avoid, or mitigate harms. Doing good and avoiding harm interact: even if the intention is to design for some good, failure to consider the full context of use may risk harm. Consider, for example, that cell phone location tracking can allow family members to find wandering loved ones with dementia but also allow abusive spouses to find their victims.

Scholarship in the social and behavioral sciences provides methods for identifying ways that technologies interact with and affect people, their interactions, and their communities. Ethical issues arise not only at the personal level, but also for communities, and computing research and technology invariably impact not only particular people but

---

[7] See, for example, H. Nissenbaum, 2001, "How Computer Systems Embody Values," *Computer* 34(3):120-129; R. Kling, 1978, "Value Conflicts and Social Choice in Electronic Funds Transfer System Developments," *Communications of the ACM* 21(8):642-657; B. Friedman and D.G. Hendry, 2019, *Value Sensitive Design Shaping Technology with Moral Imagination*, MIT Press, Cambridge, MA.

groups and societies. For example, although social networks were envisioned as bringing people together, experience has shown that they can foster divisive environments. Insights and frameworks from the social sciences thus play key roles in understanding the nature of responsible computing research, and mechanisms to ensure it.

Chapter 2 elaborates on these points and provides conceptual frameworks for considering ethical and societal issues related to computing research.

1.5 SOURCES OF ETHICAL AND SOCIETAL IMPACT CHALLENGES

As discussed at greater length in Chapter 3, the ethical and societal impact challenges raised by computing technologies arise from various sources. For example, some challenges derive from people, societies, and the functioning of governmental and other organizations—arising, for example, from conflicting values and goals of different stakeholders. Others reflect values and social structures that counted as normal or typical in the culture at earlier points in history but no longer apply. Still others result from the ways existing societal and institutional arrangements collect and use data and ways technologies are deployed. Others reflect externalities such as the environmental impacts of the energy consumed by computing systems. Some ethically or societally adverse outcomes are the result of insufficient engagement with users and other stakeholders and inadequate attention to existing social relationships or institutional structures and practices, or the lack of use of best practices in design. In almost all cases, these issues do not result from deliberately unethical behavior on the part of computing researchers, but rather from such factors as a lack of knowledge or misaligned incentives.

In addition to such societally rooted challenges, there are challenges that arise in the process of implementation or deployment. For instance, mission or function creep may occur when a technology developed for one application is applied to a new problem or in a new context for which it is inadequate, inappropriate, or poorly considered. It is thus important when reporting research results for researchers to clearly present not only the contributions of their research, but also the contexts in which it was performed and in which the results were tested and the limitations of which those using it in those or other contexts should be aware. Researchers also need to take reasonable steps, including following best practices for design and systems development articulated in this report's recommendations to anticipate other possible uses in their research and augment their work to address or at least identify potential concerns.

Computing researchers have certain obligations with respect to these challenges; other obligations necessarily fall on others. Researchers' responsibilities arise from their work being foundational—the first step in new technologies entering the world—and from their work having limitations that are important to identify and explain. In cases in which challenges stem from poor decisions by those deploying the technologies—for example, decisions that do not appropriately trade off considerations of efficiency or accountability with other social values—mitigating certain harms will involve technology businesses behaving differently and require governments to regulate. In such situations, researchers may still have a role to play as they could help illuminate trade-offs and limitations and champion (other) societal values. In many cases, this work will involve collaboration with ethicists and social and behavioral scientists.

1.6 A BRIEF HISTORY OF CONCERNS

Attention to the ethical and societal impact challenges posed by computing technologies dates back to the earliest days of computers. One of the earliest works to identify societal and ethical issues is Norbert Wiener's 1950 book *Human Use of Human Beings: Cybernetics and Society,* which drew attention to both the benefits to society of automation as well as the risks of overreliance. In 1972, SRI International researcher Charles Rosen, as part of proposing a research program to advance automation technology, called for productivity to be redefined to "include such major factors as the quality of life of workers and the quality of products, consistent with the desires and expectations of the general public,"[8] an early harbinger of concerns about the impacts of automation and computing culture that persist today. Another early example of attention to ethical issues is in Joseph Weizenbaum's *Computer Power and Human Reason: From Judgement to Calculation* (1976), which distinguishes between deciding (something that can be computed) and choice (which requires judgment) and highlights such human qualities as compassion and wisdom that computers lack.

Other fields have confronted burgeoning societal and ethical implications of their research during this span of time. For instance, in biomedicine, the 1975 Asilomar Conference grappled with the public health and ecological implications of the then-new recombinant DNA technology, following calls for a voluntary moratorium on its use. The conference concluded that research should proceed only under strict guidelines. A few years later, the 1979 Belmont Report[9] from a U.S. national commission articulated three principles for protecting human subjects in biomedical behavioral research—respect for

---

[8] C.A. Rosen, 1972, "Robots, Productivity and Quality," *ACM '72: Proceedings of the ACM Annual Conference* 1(August):47-57, https://doi.org/10.1145/800193.805821.

[9] Department of Health, Education, and Welfare, 1979, "The Belmont Report," https://www.hhs.gov/ohrp/regulations-and-policy/belmont-report/read-the-belmont-report/index.html.

persons, beneficence, and justice. Starting in the late 1980s, the Human Genome Project of the National Institutes of Health and the Department of Energy set aside 3 percent of its research budget for the study of the ethical, legal, and societal implications of the knowledge gained from the mapping and sequencing of the human genome. The interdisciplinary field of inquiry that would come to be known as science and technology studies (or science, technology, and society studies) also started to take shape during this period. Questioning technological determinism (the view that technological advances determine the development of cultural values and social structure), it emphasizes understanding the development of technology in its social and historical context. Subsequent scholarship emphasized that technologies, including computing technologies, should not be viewed as value neutral.

Also observed decades ago were ways computing technology differed from prior technology revolutions. In a 1985 essay, James Moor cited the "logical malleability" of computers—they "can be shaped and molded to do any activity that can be characterized in terms of inputs, outputs, and connectivity logical operations"—and anticipated that "in the coming decades many human activities and social institutions will be transformed by computer technology and that this transforming effect of computerization will raise a wide range of issues for computer ethics."[10] The 1980s also saw the founding of Computer Professionals for Social Responsibility (CPSR), a nongovernmental organization that focused in its early days on the risks posed by the growing use of software for military applications such as the Strategic Defense Initiative. CPSR's agenda soon broadened to look at issues that remain salient today: privacy and civil liberties, participatory design in the workplace, election systems, and encryption policy.[11]

From the 1980s onward, the use of computing has evolved markedly from use only by experts to use by nearly everyone. The 1990s saw a shift from primarily individual use (outside of some workplaces and institutions that were early adopters of computer networks) to highly interconnected use in which people's online activities connect with different people and systems. Further change came in the 2000s with the introduction of smartphones and other mobile technologies (which, with falling prices have spread around the globe) and the growing embedding of computing technologies into the physical world. In just a few decades, computing technology has become a primary means by which people interact, a primary source of functionality and value in engineered systems, and an underpinning of every sector of the economy. This radical change has engendered a whole range of new ethical and societal challenges.

---

[10] J.H. Moor, 1985, "What Is Computer Ethics?" *Metaphilosophy* 16(4):266-275, http://dx.doi.org/10.1111/j.1467-9973.1985.tb00173.x.
[11] Computer Professionals for Social Responsibility, 2005, "CPSR History," http://cpsr.org/about/history.

The development of the Internet and Web led to early concerns with and responses to abuse and manipulation of network communications. Researchers helped identify and combat with some success such attacks as network intrusions, email spam, phishing, advertising click spam, and Web spam. However, these efforts arguably did not represent enough cumulative effort given the breadth and depth of impacts being experienced today. At the same time, for much of the history of computing, the public, policymakers, and members of the computing research community have for the most part tended to emphasize positive societal and economic impacts.

Computing technology's spread has raised new issues and spurred growing recognition of the ethical and societal impacts that arise from computing research and technologies. One manifestation of this change is that universities are exploring new ways of incorporating ethics into computing courses and curricula.[12] Another is that civil society organizations, other outside observers, and even former employees are calling attention to value trade-offs being made by industry that they characterize as harmful to society. A third is a blossoming of efforts to address some of these challenges. The discussion in Chapters 2 to 4 include references to many of these efforts.

Furthermore, the National Defense Authorization Act for Fiscal Year 2021 (P.L. 116-283) signals congressional interest in the ethical implications of research in artificial intelligence. Section 5401 states that it is the sense of Congress that

> (A) a number of emerging areas of research, including artificial intelligence, have potential ethical, social, safety, and security risks that might be apparent as early as the basic research stage and (B) the incorporation of ethical, social, safety, and security considerations into the research design and review process for federal awards may help mitigate potential harms before they happen.

In recent years, various governments, companies, and other institutions have adopted different sets of ethical principles. These signal awareness of these issues although many have not yet been backed up by concrete actions.

---

[12] AI Index Steering Committee, 2021, *Artificial Intelligence Index Report 2021*, Stanford University Human-Centered Artificial Intelligence, CA, p. 134, https://hai.stanford.edu/research/ai-index-2021; Mozilla Foundation, Responsible Computer Science Challenge Winners, https://foundation.mozilla.org/en/what-we-fund/awards/responsible-computer-science-challenge/winners.

1.7 CHARACTERISTICS OF RESPONSIBLE COMPUTING IN LIGHT OF THE UBIQUITY OF COMPUTING TECHNOLOGIES

For computing research to be responsible, it needs to be ethical and adhere to societal values and norms.[13] As will be discussed further in Chapter 2, computing researchers are not free to choose norms—that is a societal prerogative—but need to be knowledgeable of them and take them into account in their research.

Computing research must thus consider and take into account its potential societal impacts, especially now that computing technology is present throughout the daily lives of individuals, communities (of work and of play), and society. Society also expects computing technologies to be trustworthy, transparent, and accessible and designed in ways that ensure that users can understand and control what the technologies are doing on their behalf. These expectations have become ever more important with the increased complexity and scale of today's computing systems.

One might think these expectations apply only to computing systems research, but they apply as well to theoretical work. For example, choices made by researchers to improve the performance of a matching algorithm can raise significant societal impact and ethical questions when that algorithm is used, say to optimize kidney transplant organ exchange. For example, an objective function used to improve the efficiency of transplant matches might turn out to favor individuals from some groups over others depending on the way the matching algorithm resolves a tie. Indeed, ethical and societal impact questions are not just arising in theory papers. An entire conference devoted to the topic, the Symposium on Foundations of Responsible Computing, has been held annually since 2020.

There are many ways that computing researchers can take into account the broader context:

- Sufficiently deep grounding in the intended application domains. Where computing research is concerned with use in a particular application or context, for example, in such sectors as health care, education, or transportation, computing research will benefit from researchers at a minimum engaging with experts in that application area and potentially including such experts as part of the research team.

---

[13] Although some current norms may need to be changed, that is a societal responsibility. Computing researchers can, of course, with their research decide to support such changes. Note also that it is possible for societal norms to be unethical. Such problems are for society to sort out, but computing researchers can call attention to such conflicts.

- Seriously considering potential uses and application domains beyond those originally contemplated, including both beneficial and problematic uses. Although it is not possible to predict all future uses, it is possible to increase the probability of finding the most likely ones by engaging with people knowledgeable about the original application domain and using well-established design methods. For example, a predictive algorithm used in pretrial detention decisions may not be appropriate for making parole decisions. A related issue is to anticipate and warn against mission creep, where a technology developed for a narrow use is subsequently used in a much broader context (as in the use of cookies referenced above).
- Engaging with or otherwise considering the perspectives of all the stakeholders in the intended context of use, including not only direct users but others who will be affected by its use and considering the power differences among the stakeholders including the researchers.
- Collaborating with scholars who have expertise in the social and behavioral sciences, and the humanities, most notably in ethics and ethical reasoning.

Relevant areas of expertise include the humanities, social and behavioral sciences, and ethical reasoning as well as any particular domain of intended use (e.g., health care). They may also include other possible domains of use to help identify potential other deployment settings. The report and its recommendations do not anticipate that computing researchers will become scholars or experts in any of these fields or domains. Rather they can successfully incorporate such expertise into their projects by collaborating with people with expertise in these areas. Doing so effectively entails that computing researchers acquire knowledge in some areas of the social and behavioral sciences and humanities and also that humanities and social science scholars understand key computing concepts. Furthermore, to successfully incorporate the requisite expertise into computing research projects may not only require the involvement of new kinds of expertise beyond that traditionally involved in computing research but also new kinds of projects that effectively leverage this new expertise. The recommendations also describe steps research institutions and research sponsors need to take to facilitate and support such efforts.

1.8 STUDY APPROACH

Responsibilities for computing technologies' effects on people and society progress from research to product and service deployment. Some responsibilities rest with researchers; for example, in how they scope and structure projects, the diversity of perspective and expertise they engage, and how they report research results, including limitations or caveats. Other responsibilities rest with industry; for example, in what technologies are deployed, how, and for whom. Still other responsibilities rest with the government, which has responsibility for setting policy objectives, writing legislation, incentivizing desired behaviors, and formulating needed regulations.

The analysis and recommendations in this report are primarily aimed at the computing research ecosystem comprising computer researchers, the computing research community, the scientific and professional societies in which they participate, other scholarly publishers, the public- and private-sector agencies and organizations that sponsor computing research, and the public- and private-sector institutions that perform computing research. The committee has attempted to formulate recommendations that will work for all research producing institutions—that is, for small colleges, public colleges and universities, and private universities, and for industry and government research as well as academic organizations. It has fashioned them with the understanding that the institutional resources for carrying out the recommendations will vary, and that some actors may need to step in to support others, such as scientific and professional societies assisting less-resourced institutions with accessing scholars with the requisite expertise on societal and ethical aspects of research proposals and activities. Because there is little empirical data on the effectiveness of any approaches to responsible computing research, these recommendations have been developed primarily by considering leverage points in the research ecosystem, early efforts that appear promising, and expertise provided by social scientists and ethicists who served on the study committee and made presentations to the committee. Several recommendations also address the need for empirical evaluation and possible adaptation or revision of some recommendations based on experience implementing them.

The recommendations are also designed for use by government funding agencies as well as industry and philanthropic research sponsors. In keeping with its statement of task, this report does not provide recommendations for government regulation of computing technologies including corporate computing research, but it does discuss ways that the computing research community can help inform government action in this space.

The report describes various ways that addressing diversity, equity, and inclusion (DEI) in the computing research community is critical to ensuring responsible computing research. (The same holds true for the other disciplines and application domains involved in the research.) The committee considers DEI a cross-cutting issue, and discussions of it permeated the committee's analysis and informed its recommendations. Although the report does not offer separate recommendations on DEI, these considerations are reflected throughout many of the recommendations.

By defining practices by which more ethical and societal adverse outcomes can be caught than at present and better steps taken to mitigate or eliminate potential harms to individuals and society, the report's recommendations aim to foster responsible computing. They cannot, of course, ensure that every potential ethical or societal problem in the computing research ecosystem will be recognized and addressed. In some cases, the recommendations constrain research explicitly and call for direct action by computing researchers, while in others they speak to the roles researchers have in assisting those deploying computing research outcomes and technologies to use them in ways that take into account their limitations as well as strengths.

2
Theoretical Foundations from Ethical and Social Science Frameworks

This chapter provides theoretical foundations for identifying the roots of ethical challenges and sources of problematic societal impacts in computing research, which are described in Chapter 3, and for recommendations for addressing them, which are presented in Chapter 4. It describes core ethical concepts (Section 2.1) and fundamental ideas from social and behavioral sciences (Section 2.2). These foundations enable identifying, understanding, and thus better addressing ethical and societal dilemmas that arise with computing research and in the technologies it engenders. The chapter aims to give computer scientists and engineers, most of whom are neither philosophers nor social scientists, a basic understanding of the major ideas in ethics and social sciences that will assist them in carrying out responsible computing research. Necessarily, given its brief nature, the chapter's presentations are not in-depth in either area of scholarship. Importantly, this report does not assume, nor expect, that computer scientists and engineers will become experts in these areas of scholarship. Rather, the goal, both of this section of the report and through the recommendations, is to enable them to participate in meaningful collaborations with scholars in these fields, so that their computing research may be better informed and more responsible to societal needs.

The social and behavioral sciences provide methods for identifying the morally relevant actors, environments, and interactions in a sociotechnical system; ethical reasoning provides a calculus for understanding how to resolve competing moral tensions involving those actors, environments, and interactions. The theoretical foundations presented in this chapter can thus support the computing research community in identifying and making informed decisions about ethical and societal impact challenges that arise in their research. They provide a basis for determining ways to adapt for responsible computing the processes of design, development, deployment, evaluation, and monitoring of computing research and thus help guide responsible downstream use of computing research in building the many products that are reshaping daily life. In particular, scholars with expertise in these areas can assist computing researchers in designing research projects that adequately meet societal constraints, norms, and needs.

To address the ethical and societal impact challenges discussed in this chapter, computing researchers need to be able to envision alternative ethical values and trade-offs among them as well as alternative socio-technical contexts. Scholarship in the field of design provides methods and techniques for effectively doing such envisioning. These methods and techniques are frequently deployed in human–computer interaction research and are now typically included in courses on this topic in computer and information science curricula. The report discusses design in Sections 3.3.1 and 3.4.4.

2.1 THE VALUE AND SCOPE OF ETHICS

In recent years, sets of principles aimed at guiding those engaged in the development and deployment of artificial intelligence (AI) systems toward ethical and socially beneficial outcomes have proliferated.[1] To advance responsible computing research in general, one might consider taking these principles as a baseline, and expanding on them to encompass other areas of computing, such as cybersecurity or software engineering. Principles may be a natural starting place for developing recommendations for responsible computing research, and indeed, many such principles make meaningful contributions to society's continuing deliberation about the present and future of computing. Nonetheless, they are insufficient in themselves because they are both often relatively divorced from practice and tend to be presented absent sufficient explanations of their underlying assumptions or origin in ethical reasoning. For example, a principle that says that a system must be "governable" or an algorithm's results "interpretable" provides little, if any, guidance about ways to develop or test for these properties. Similarly, a principle that says that a research project or product should be "respectful of human dignity" is unlikely to make a practical difference in isolation. Without shedding light on core assumptions about the fundamental ethical concepts, social theories, and humanist and social scientific factors that underlie these principles, researchers lack guidance on how to interpret, critique, and apply them in practice. This chapter focuses instead on presenting fundamental ethical concepts, the very concepts from which such principles

---

[1] AI Index Steering Committee, 2022, "AI Policy and Governance," Chapter 5 in *The AI Index 2021 Annual Report*, Stanford University Human-Centered AI Institute, CA, https://aiindex.stanford.edu/report.

arise, but, more importantly, concepts which support the practical reasoning responsible computing research requires.

Ethics provides tools for the moral evaluation of behaviors, institutions, and social structures. This section focuses on evaluation of behaviors, and Section 2.2 examines the roles of institutions and social structures. The tools for evaluating behaviors provide building blocks for ethical evaluation of computing research, including a language and concepts in which to express a set of baseline commitments against which to assess research. When assessing behaviors, one must distinguish between the moral evaluation of acts and that of agents. The first concerns what people ought to do; it aims to identify the right act to choose. The second concerns practices of moral blame and praise, and aims to identify who is responsible, and to what degree, when a morally right or wrong act is performed. Both these questions are important: achieving responsible computing research requires not only determining whether an action (e.g., a design choice) was responsible, but also determining who or what is responsible for that action. Responsibility comes in degrees depending on the nature of the researcher's contribution and the source of any harm that ensues from the system.

In addition to determining whether an act is wrong, ethical evaluation also requires determining why the act is wrong and how seriously wrong it is. This further evaluation requires examination of values that are put into play by the computing researchers' decisions, and determination of the extent to which their decisions undermine (or serve) those values. Institutional and social structures can also promote or undermine relevant values; for example, through promotion and tenure evaluations, or review criteria at conferences. Ethical evaluations often require examination of many mechanisms that impact relevant values, as well as potential conflicts between different values.

Consider, for example, evaluations of decisions about how to deploy potentially privacy-invasive computing technologies in order to support public health responses during the COVID-19 pandemic. The use of smartphones to support contact tracing by creating Bluetooth "handshakes" with other devices within a given range for a given period might help advance public health goals, especially if such information is integrated into manual contact tracing. But sharing information about people's "social graph" with a centralized health authority (as well as storing it on a device) can raise real privacy concerns. Understanding the contributions each makes to supporting values such as autonomy, well-being, and the legitimate exercise of power can help structure a well-reasoned evaluation of this potential trade-off.

More generally, ethical evaluation fundamentally requires weighing multiple values, and both the values and weightings of them are domains of intense disagreement. This report neither adopts a particular value system nor provides a complete decision procedure for resolving the conflicts and tensions that inevitably arise. The concepts it articulates cannot be used to derive an "ethical checklist," either for computing artifacts or for computing researchers. Indeed, no simple checklist could suffice for determining if research is responsible; there is no mechanical procedure that can spare researchers from having to think about the values they embed in their research and the trade-offs they make in doing so.

2.1.1 From Ethical Theories to Ethical Values

Philosophers have developed a number of distinct ethical theories, each of which may be mobilized to determine whether an act, in a context, is morally required, permissible (but not required), or impermissible. These moral theories differ less with respect to the practical verdicts they endorse and more with respect to how they explain those practical verdicts[2]—a topic beyond the scope of this report. This report's engaged ethics approach aims at being useful for the purposes of guiding responsible computing research: instead of applying some canonical, abstract ethical theory, it starts from an engagement with responsible computing issues and aims to identify the ethical concepts and reasoning that can be used to approach resolving them, sometimes yielding new theory.[3]

This section thus focuses on the fundamental building blocks of moral theories, namely, ethical values.[4] The authors expect most reasonable moral theories to agree that these values are important, even though the theories may not agree on the precise details of the values or the ways to deploy them in an ethical argument.[5]

The space of plausible values is vast. One useful contribution of moral theories is to provide a shorthand for thinking about those values and their structure. So, this report adopts a pragmatic distinction to facilitate ethical analysis of responsible computing research: the distinction between intrinsic and instrumental values.[6] *Intrinsic values are things that matter in themselves. Instrumental values are things that matter because they help us to realize intrinsic values.* Intrinsic values are typically more abstract, general, and

---

[2] D.W. Portmore, 2009, "Consequentializing," *Philosophy Compass* 4:329-347, https://doi.org/10.1111/j.1747-9991.2009.00198.x.

[3] A. Cribb, 2010, "Translational Ethics?: The Theory–Practice Gap in Medical Ethics," *Journal of Medical Ethics* 36(4):207-210; J. Johnstone, 2007, "Technology as Empowerment: A Capability Approach to Computer Ethics," *Ethics and Information Technology* 9(1):73-87; D. Danks, 2021, "Digital Ethics as Translational Ethics," Pp. 1-15 in *Applied Ethics in a Digital World* (I. Vasiliu-Feltes and J. Thomason, eds.), IGI Global, Hershey, PA.

[4] J. Raz, 1999, *Engaging Reason: On the Theory of Value and Action.* Oxford University Press, United Kingdom; M. Schroeder, 2021, "Value Theory," In The Stanford Encyclopedia of Philosophy, Fall 2021 ed. (E.N. Zalta, ed.), https://plato.stanford.edu/archives/fall2021/entries/value-theory.

[5] In particular, different substantive ethical theories may disagree about the exact scope of particular values (e.g., does privacy apply to email communications?) or the relative weights of the values (e.g., should transparency or privacy be valued more when those come into conflict?). Historically, different cultures have often endorsed different substantive theories, particularly with regard to how different values are weighted or traded off against one another. Despite these differences, these values can provide valuable building blocks for substantive theories.

[6] M.J. Zimmerman and B. Bradley, 2019, "Intrinsic vs. Extrinsic Value," In *The Stanford Encyclopedia of Philosophy,* Spring 2019 ed. (E.N. Zalta, ed.), https://plato.stanford.edu/archives/spr2019/entries/value-intrinsic-extrinsic.

as a result fewer in number. Instrumental values are typically more applied, specific, and as such are potentially infinite.

This division can help us navigate substantive moral disagreement. Ethics scholarship generally agrees that the concepts described here as intrinsic values matter, but often disagrees about how much each intrinsic value matters or how to balance them. Because the importance of each instrumental value depends on its connection with various intrinsic values, ethical debates can be conducted in terms of the smaller set of agreed-upon intrinsic values. For example, suppose one has apparently conflicting instrumental values, but both matter because they help realize the same intrinsic values. In that case, one can translate the debate into the intrinsic values to provide a manageable currency in which to express, and potentially resolve, this apparent conflict. More generally, this division enables one to reduce the potentially infinite list of possible instrumental values to the more tractable set of intrinsic ones.

The lists below focus on relatively canonical intrinsic values and on instrumental values that are relevant for computing research. One challenge is that different philosophical traditions may use different terms for some values, even when the core value (and related concepts) is shared. Moreover, there is some philosophical disagreement about the exact list of intrinsic values, though agreement about the general content of the list. Some also suggest using rights to establish ethical foundations, but this is problematic.[7] And of course, beyond this philosophical disagreement, there is considerable substantive moral disagreement across cultures, places, and times. The presence of pervasive disagreement may be daunting to computing researchers who are disposed to seek determinate solutions to quantifiable problems and may lead some to seek to avoid ethical considerations entirely. Although it is impossible to eliminate this pervasive moral disagreement, it is possible to provide concepts that enable computing researchers and others to better understand those disagreements. By identifying a set of widely (though not universally) shared intrinsic values and illustrating how they are served by instrumental values specific to computing, this report offers computing researchers concepts with which to structure and understand both their own moral intuitions, and the inevitable

---

[7] One approach to addressing ethical questions in computing research would be to rely almost exclusively on ideas of rights, perhaps rooted in domestic or international law. At least in principle, for example, virtually all countries in the world accept the idea of human rights. Although this report does not dismiss the importance of rights as a rhetorical framing for public discussions and debates about the implications of computing research, it relies on a somewhat different approach for both theoretical and practical reasons. First, the notion that rights provide foundations is itself illusory, as rights must be grounded in values that are important enough in a context to generate a duty that someone owes to the holder of the right. Second, the appearance of agreement on universal human rights is superficial and depends on the articulation of those rights being vague and general. Efforts to make universal human rights more precise inevitably leads to the same disagreements as with any other ethical concept. Third, rights are not useful for injuries that are significant only in the aggregate. Although this report is not grounded in rights, there are contexts in which such language is valuable, such as giving a name to duties to one another that might otherwise go unrecognized.

moral disagreements that they will confront, when assessing the sources of ethical and societal challenges discussed in Chapter 3 and the recommendations in Chapter 4.

2.1.2 Intrinsic Ethical Values

- *Autonomy and freedom*—Individuals have beliefs, plans, and goals, and autonomy is the ability to act on those beliefs by formulating plans to achieve goals. Different people's autonomy, as well as their perceptions of autonomy (or its absence), can obviously come into conflict, and so philosophers have advanced more and less substantive conceptions of freedom and autonomy, including explanations of when perceived autonomy is relevant. At one extreme, so-called negative freedom consists simply in being free from interference by others. At the other extreme, so-called positive freedom consists in being able to formulate authentic beliefs and goals, and actually realize those goals. Autonomy as positive freedom also presupposes having a sufficient range of good options to choose from.[8]

- *Well-being (material and non-material)*—People's well-being is an intrinsic value, as all people have an interest in achieving suitable levels of physical and psychological functioning. Material well-being requires access to sufficient sustenance, water, shelter, and so forth. Non-material (psychological and social) well-being is similarly an intrinsic value, described by some philosophers as "the social bases of self-respect." This type of well-being can be negatively impacted by, for example, representational harms by algorithmic systems that reproduce racist tropes. This value naturally translates into a right for access to a basic level of psychological and social health, even if people autonomously choose not to use that access.

- *Relational and material equality*—Relational equality refers to people standing in equal social relations to one another such that, for example, one person is not unilaterally vulnerable to the other, or where each is an equal partner in decision-making. Many regard this type of equality as an instance of mutual regard for human dignity. The ethical wrong of exploitation (of other people) is closely associated with the intrinsic value of relational equality, though exploitative acts and policies can impact other intrinsic values as well.

---

[8] There have been many efforts to develop viable universal, inclusive, and trans-cultural conceptions of the substantive bases for "real" autonomy, in contrast with the relatively easier task of evaluating comparisons in the extent of autonomy experienced by particular persons, or merely assessing subjective perceptions of autonomy among individuals or groups. For example, it is widely accepted in human rights law that women ought not have less autonomy than men, but that comparative claim does not establish criteria for determining whether persons in general objectively have "sufficient" autonomy, nor does it necessarily resolve conceptual questions about whether autonomy means the same thing in different contexts. These issues can be important for topics like surveillance that touch all members of a society equally. Computing researchers whose efforts implicate such questions are strongly encouraged to learn more about these complex issues.

- *Justice and legitimate power*—Equality is a fundamentally comparative value—it concerns how people stand in relation to each other, or how their material well-being compares with others. There is also a distinct value of justice that is noncomparative. Justice focuses on ensuring that people receive what they are due, which may require the use of legitimate power. In modern societies, people's ability to live according to their intrinsic values (and be protected from others' actions) can require an authority to exercise power and thereby resolve disputes, protect the vulnerable, and enact collective self-determination. It is intrinsically valuable that this power be exercised legitimately—in ways that are limited, and impartial, and are properly authorized by the community whom that authority represents. Of course, in many contexts, one may also find significant instrumental value in justice and legitimate exercise of power.
- *Collective self-determination*—Just as some forms of individual autonomy may be intrinsically valuable, so the self-determination of groups and collectives can have intrinsic value. One's life plans frequently depend on coordinated planned action with others, and so too there is value in one's group being able to autonomously pursue legitimate plans. This value does not entail that collectives have values above those of their members, but only that individuals can find intrinsic value in the success of their groups.
- *A thriving natural environment*—The natural world arguably has moral status on its own, independently of our human interests. As such, there are arguments that a thriving natural environment is intrinsically valuable analogous to human thriving being intrinsically valuable. Of course, the environment is also instrumentally valuable in the ways that it enables us (and other moral beings) to thrive.

2.1.3 Instrumental Ethical Values

Instrumental values are ethically important because they contribute to the realization of (or capability to realize) intrinsic values. Instrumental values thus tend to be a more heterogeneous collection than intrinsic values, as their ability to contribute to people's intrinsic values will depend partly on the particular context, environment, and agents. Instrumental values are sometimes not actually valuable, as pursuit of them might not lead, for that agent in that context, to any intrinsic value; they must be assessed in context-sensitive ways. The instrumental values listed below are ones frequently raised in conjunction with computing research. Some expectations for responsible computing research are themselves an amalgamation of instrumental values. For example, such values as privacy, trust, and transparency are relevant to the value of non-exploitative participation or use—that is, to ensure that an individual's participation, activity, or data is used in ways that the individual understands and agrees to.

- *Privacy*—The ethical value of privacy arises in many different spheres, not only those to do with computing research and product deployment. In the computing context, analyses of privacy all focus on its importance for protecting some other important value, though conceptions of privacy differ across individuals, cultures, and contexts. Privacy can provide protection against manipulation or coercion (supporting autonomy); enable formation of intimate relationships that partly depend on the secrets shared with loved ones (supporting psychological well-being); protect against overreaching and illegitimate governments (supporting collective self-determination and legitimate power); or yield other support depending on the computing use and context.
- *Avoidance of unjust bias*—Unjustly biased systems potentially undermine material well-being, non-material well-being, justice, autonomy, and collective self-determination for those against whom they are biased, while also often undermining relational equality. For example, university admissions systems that are biased against people with working-class backgrounds can result in the harm of denied opportunities, and biased medical resource allocation algorithms can divert resources for medical care away from needy but historically disadvantaged populations. In general, unjustly biased systems perform worse for members of historically disadvantaged groups than for members of historically advantaged groups, without an ethically defensible reason for this bias. As a result, these systems can also damage non-material well-being even when deployed in contexts with lower stakes than these.
- *Fairness*—Although there are many conceptions of fairness, most imply that one should use the same kinds of ethically defensible reasons for all decisions of a particular type. For example, one might use only an applicant's income to determine whether to approve a loan, as someone's income is clearly relevant to their ability to repay the loan. The ethical value of fairness is not the same as the absence of bias. As this example shows, fair systems can sometimes be biased (depending on the broader context); conversely, unbiased systems can nonetheless be unfair (as when a monopolistic company unfairly charges higher prices to everyone). There have recently been a number of proposed statistical measures of fairness, and while those might provide signals or guidance about potential (un)fairness, it is important to recognize that they are not constitutive of it.

- *Trust and trustworthiness*—At a high level, trust involves someone becoming vulnerable in certain respects because of (justified) expectations about the person or tool being trusted; trustworthiness is the property of a person or tool that makes such trust reasonable. Trust thus enables people to do or realize much more when that trust is appropriately placed. For example, a trustworthy computing system could be valuable because it maintains data integrity, or learns from incorrect predictions, or otherwise supports intrinsic values such as material well-being or autonomy.
- *Verifiability*—People understandably have an interest in knowing that a computational system will function correctly so that they can use or engage with it appropriately. That is, people instrumentally value being able to foresee a system's behavior, because that knowledge enables them to use the system to advance other values. The importance of knowing that a system will behave correctly is closely aligned with computing research on (formal) verification.
- *Assurance*—People also value having reasons for believing that a system, or another human, will behave in expected ways. Reason-giving is a ubiquitous feature of ethical debate and discussion, often serving to excuse perceived ethical lapses or errors; we do not just value knowing what others will do, but also why they will. This information is particularly (instrumentally) valuable because it supports successful interaction in new contexts. Computing research on system assurance similarly focuses on the value of providing reasons to expect that the system will behave appropriately, particularly in complex environments.
- *Explainability, interpretability, and intelligibility*—These concepts are grouped together as they have all been proposed as ways to promote understanding of increasingly complex computational systems, and thus to support meaningful deliberation, oversight, or use of these systems. There are no widely shared definitions of these terms—for instance, one person's explainability is another's interpretability—but they have a shared conceptual core of increased understanding, whether about predictions, control, potential improvements, or transfer. Improved understanding is clearly ethically valuable in many cases, but always because of what it enables—increased autonomy, better outcomes, lower risks, greater security, accountability in the exercise of power, and so forth.
- *Safety*—In computing contexts, safety primarily concerns the design and structure of the system, and whether it will behave appropriately such that it does not kill, injure, harm, or otherwise endanger the well-being of users or other

individuals.[9] This focus is closely related to the broader ethical value of being able to act as desired without active threat. It thereby supports the intrinsic values of well-being and autonomy, either directly or through changes or preservation of the natural environment.

- *Security*—The related concept of security has a specific meaning in computing research focusing on performance of real-world implementations (with potential adversaries), and this applied concept is rooted in a deeper ethical concept, according to which the intrinsic values described above should be enjoyed securely—that is, without worry that significant threats or harms might arise. Protection against novel or additional threats makes actual harm less likely, and so enables people to better realize their intrinsic values.

- *Transparency*—In computing, this notion encompasses a broad range of goals, including algorithmic transparency, clarity, and specificity about system capabilities, and information about how one's data are used. Techniques that support explainability, interpretability, and intelligibility are one way to increase transparency. These goals can be valuable to ensure appropriate use or understanding of computing systems for people to realize their intrinsic values.

- *Inclusiveness and diversity*—Inclusion of a range of diverse perspectives is frequently emphasized as an important value in computing research contexts, including by this report (see Sections 1.6 and 3.1). There is growing recognition across many sectors—government, industry, academia—that diverse and inclusive teams make better decisions. Increasing diversity and inclusion—in research teams as well as in the stakeholders who are consulted in carrying out the research—also supports relational equality and collective self-determination, when everyone feels empowered to contribute to important decisions that affect their community. Non-material well-being—the social bases of self-respect—is also enhanced by removing barriers for underrepresented groups and maintaining environments that are conducive to their participation.

Ethical challenges often involve conflicts between values. For example, decisions about surveillance systems including facial recognition almost always involve trade-offs between increased security and increased privacy, as when decision-makers must decide how many cameras to deploy, or indeed whether to adopt facial recognition algorithms at all. In this example, computing research can potentially help change this trade-off (e.g., through privacy techniques embedded in the cameras themselves), but not fully eliminate it. Or consider the trade-offs that arise when only some people benefit while

---

[9] M. Bishop, 2019, *Computer Security,* 2nd ed., Addison-Wesley, Boston, MA, p. 630.

others are potentially harmed: for example, software engineers are often pressured to rush products that have not been fully tested, thereby pitting the values of some end-users (e.g., their material well-being if the system fails) against the values of other end-users (e.g., their autonomy to be free to use the product) against the values of the employees (e.g., to advance the company in personally beneficial ways). The concepts, ideas, and framework articulated here provide the resources to engage in careful analysis and decision-making about cases like these, even if no simple checklist or decision procedure is possible. Section 3.1.1 further explores the challenges in reconciling conflicting values and goals of stakeholders.

2.2 THE POWER OF A SOCIOTECHNICAL PERSPECTIVE

The technological artifacts that computing research creates—from algorithms and other computational methods to networked systems—participate in an increasingly complex and highly interconnected ecosystem of people, institutions, laws, and societies. It is computing's participation in this social ecosystem and, consequently, its far-reaching societal impact that give rise to the ethical challenges and questions of responsibilities increasingly posed to computing research. The *sociotechnical approach*[10] explained in this section provides an important framework for the computing research community in its pursuit of understanding ways to identify and address these challenges and calls for greater accountability. The term *sociotechnical* conveys more than the fact that computing systems have users. It highlights that people—individually, in families, at work, in communities, as members of society, and so on—are interacting with and affected by computational systems, tying humanity to the material, technical, and social worlds created by computing research.[11]

A sociotechnical approach enables identifying, designing for, and tracking the benefits and risks that arise from introducing novel technologies into social worlds. It draws on social theories and social scientific methodologies, and empirical observations that enable the development of hypotheses about the ways people interact with the world

---

[10] W. Bijker and T. Pinch, 1987, "The Social Construction of Facts and Artifacts: Or How the Sociology of Science and the Sociology of Technology Might Benefit Each Other," *Social Studies of Science* 14:17-50, First published 1984 in *Social Studies of Science* 14(3):399-441; S. Sismondo, 2011, *An Introduction to Science and Technology Studies,* John Wiley & Sons, Hoboken, NJ; E.J. Hackett, O. Amsterdamska, M. Lynch, and J. Wajcman, eds., 2008, *The Handbook of Science and Technology Studies,* MIT Press, Cambridge, MA. Other relevant resources include the ACM Computer-Supported Cooperative Work and Social Computing (CSCW) Conference, which has been hosting a robust conversation about sociotechnical systems since the 1990s, the European CSCW Conference and CSCW Journal, and the field of Science and Technology Studies' flagship journal *Science, Technology and Human Values,* which has a wealth of important research about sociotechnical systems.

[11] J. Hughes, 1989, "Why Functional Programming Matters," *The Computer Journal* 32(2):98-107, https://doi.org/10.1093/comjnl/32.2.98.

around them. These theories and methods thus enable investigations of the ways people interact with computing technologies in various contexts and circumstances of use and of the complex roles that technologies play in those dynamic interactions. They are essential to revealing ways to enable computing research to be responsible in this time of widely deployed and highly networked computing systems.

The analytical methods the social and behavioral sciences use to generate meaningful insights about the world include ethnographic observation methods, in-depth interviews, survey studies, historical analysis, and simulations and experimental studies in controlled lab settings.[12] These methods have been developed and deployed by scholars in many disciplines—including anthropology, information science, education, ethnic studies, history, qualitative sociology, political science, public health, urban studies, and women and gender studies. Computing researchers cannot be expected to become experts in any one of these disciplines, let alone all of them. They can, however, collaborate with scholars in these disciplines, who can assist them in understanding and applying such nuanced concepts as society, gender, race, justice, and systemic oppression that are studied in the social sciences.

Of the many social science approaches, the discussion in this section draws mainly on insights from the interpretivist paradigm, an approach that focuses on understanding how people make sense of their everyday lived experiences and they ways those understandings shape what people do and value in the world. It provides effective methods for analyzing the relationships among science, technology, and society, enabling study of ways new technologies may be adopted by people and the organizations in which the operate. In particular, this approach draws social considerations into focus for even the most seemingly purely technical systems, and thereby shows the advantage of framing the challenges of responsible computing research as sociotechnical problems. More socially accountable technologies that support people's manifold values require the expertise of both social scientists and computing researchers.[13]

2.2.1 Sociotechnical Systems Briefly Explained

The social contexts that participate in this feedback loop include the many interpersonal, linguistic, cultural, professional, institutional, and historical experiences that shape individuals as well as their personal experiences. Technologies take on their meaning

---

[12] See, for example, J.S. Olson and W.A. Kellogg, eds., 2014, *Ways of Knowing in HCI, Vol. 2,* Springer, New York; M.Q. Patton, 2014, *Qualitative Research and Evaluation Methods: Integrating Theory and Practice,* Sage Publishing, Los Angeles, CA; U. Felt, R. Fouché, C.A. Miller, and L. Smith-Doerr, eds., 2016, *The Handbook of Science and Technology Studies,* MIT Press, Cambridge, MA; M.J. Salganik, 2019, *Bit by Bit: Social Research in the Digital Age,* Princeton University Press, NJ.

[13] G. Ropohl, 1999, "Philosophy of Socio-Technical Systems," *Society for Philosophy and Technology Quarterly Electronic Journal* 4(3):186-194, https://doi.org/10.5840/techne19994311.

and value in particular places, moments in time, among networks of people, and their physical environments. Even the simple technology of scissors was designed initially with a notion of a universal person in mind—the right-handed person. As a result, seemingly purely technical systems are not just technical, and approaches to responsible computing must grapple with the myriad ways computing research interacts with people and the social contexts they inhabit. The facial recognition example presented below illuminates and explores the sociotechnical nature of computing technologies.

Computing research not only shapes, but also is shaped by a range of values, priorities, influences, and effects. Stakeholders of various types are embedded in the social contexts of computing research, including funding agencies and academic peer reviewers and investors, who influence the deployment and use of computing research. Thus, the computing research enterprise itself participates in a sociotechnical system.

Another insight from social science scholarship is that the effects of computing research results and the products it enables are influenced by social phenomena at multiple scales.[14] Macroscale social phenomena include national laws, economic conditions, and shared political ideologies; mesoscale social phenomena include organizational cultures and institutional rules; and microscale social phenomena include interpersonal relationships and shared identities. These different scales of social phenomena interact, particularly because they might embody different core values.[15] For example, micro-level interactions (such as the treatments a patient is offered by a clinician) are shaped by meso-level cultural norms (such as whether there are patterns of systemic racism in the admissions practices of a regional hospital) and macro-level policy decisions (such as laws that restrict who can legally access health care). Individuals help shape the kinds of research questions that are asked but so do the organizational cultures and incentives that influence whose questions are pursued and what approaches are taken.[16] The interplay of these connections and social phenomena makes technical systems challenging to design, build, and deploy.

2.2.2 From Image Recognition to Facial Recognition Technologies

The importance of a sociotechnical perspective can be readily seen by considering the contrast between image recognition and facial recognition systems. From a purely computing technical perspective, it might seem that facial recognition technology is merely

---

[14] Social scientists might characterize these scales with more nuance than the simple explanations given here for the sake of those in other fields.

[15] E.L. Trist, 1981, *The Evolution of Socio-Technical Systems, Vol. 2,* Ontario Quality of Working Life Centre, Toronto.

[16] M.S. Ackerman, 1998, "Augmenting Organizational Memory: A Field Study of Answer Garden," *ACM Transactions on Information Systems* 16(3):203-224, https://doi.org/10.1145/290159.290160.

"image recognition applied to faces." However, there are stark differences between the technologies when one adopts a broader perspective.

Facial recognition technologies were built on a body of image recognition research. One important technique, the simple, elegant tool of box-bounding, a core image recognition technique, provides a mechanism for digitally drawing (and using) discrete boundaries around objects. Box-bounding makes it possible for human annotators to mark the borders between the features of an image to help computer vision methods develop the ability to distinguish among different objects, including faces. The second was the use of human work to assist the machine learning systems by labeling image recognition training data, which harnesses human work to assist computing systems, to label image recognition training data.

Four research advances enabled this second development: the capacity to collect and store images from vast troves of user-generated content scattered across the Internet (that could serve as training data); the development of platforms for effective labeling of images by people at a large scale; the increased computing power of graphics processing units; and the development of faster deep neural network algorithms.

ImageNet, a large collaborative project initiated by Fei-Fei Li at Stanford University, is one of the earliest and most well-known examples of the successful development of an image recognition model for computer vision human-labeled training data. The success of ImageNet validated a methodology for benchmarking ground truth in machine learning. With enough data, it became possible to build reliable classifiers for various types of images, including, importantly for facial recognition.

Although some researchers at the time thought about the possible consequences of image recognition systems, there were few if any in-depth sociotechnically informed investigations of the socially relevant consequences of automating computer vision of social worlds. However, the social consequences of image recognition change significantly when the images being recognized are human faces. Two components of this transition raise important ethical and societal impact issues. These issues arise as well for other non-neutral objects as cultural artifacts.

First, researchers were able to amass a collection of faceprints that were purchased, donated, or surreptitiously scraped from image stashes available online through myriad sources—everything from local newspapers to Flickr and other photo hobbyist sites. This case provides a compelling illustration of the conflicts that can arise between different instrumental values (Section 2.1)—here between low data acquisition costs for researchers and the autonomy of the subjects whose data are used. Hundreds of computer researchers, using these collections, advanced the ability of software to identify any single face with a computational projection of the mathematical likelihood that an image taken in real time matched the face in front of it. Second, combining box bounding with platforms for human labeling of images created a powerful mechanism for image classifiers—the capacity to rapidly aggregate human decisions to validate and structure large quantities of data—in particular for identifying faces in images.[17]

The collection and use specifically of facial data raise novel societal issues: What images might different people, with diverse languages, cultures, and laws, consider sensitive, profane, or private? What collection methods (beyond confirming the data) were secure? How was the privacy of the individual supplying the image protected? What if people didn't want to be part of a research project? It also raises some issues familiar from other contexts about what rights those posting their images publicly on the Web have with respect to various ways their information might be used.

This workflow also raises distinctive issues when applied to faces, particularly with regards to training of the image classifiers and the potential uses of these classifiers: What if people whose faceprints are bucketed into the same demographic category for training data would not agree with where they are placed? What are appropriate and inappropriate uses of ImageNet and face recognition software? What kinds of governance structures need to be in place to ensure appropriate use of and access to ImageNet and other data sets, or of the uses of facial recognition overall?

2.2.3 Characteristics of Sociotechnical Systems

Facial recognition technologies illustrate several key characteristics of sociotechnical systems: interactivity of social scales and technology design, divergent stakeholder values, challenges of achieving universality, the role of social historical contexts, limited predictability of future uses, values implicit in design, continuous integration and evolution, and impacts beyond the individual level. Each of these is described briefly below.

Challenges of Achieving Universality

People are more than the sum of their personal experiences and individual attributes. They share group identities as well as linguistic and cultural traditions, and they navigate a world of laws and economic conditions. Models of individual human cognition and behavior on which computing research often relies tend to flatten or disregard such social phenomena. People who design a technical system may even start with themselves as the typical user without explicitly thinking through how much or how little they resemble the breadth and depth of human experience. The people represented in data sets may be only those who use a specific system, and thus not be representative of any larger span of humanity than that population. For instance, the images in a facial

---

[17] M.L. Gray and S. Suri, 2019, *Ghost Work: How to Stop Silicon Valley from Building a New Global Underclass*, Houghton Mifflin Harcourt, Boston and New York.

recognition system represent only people like those who have uploaded images. Analogous problems arise for systems designed with an imagined typical user (or, abstractly, "anyone") rather than careful attention to engaging the full range of stakeholders.

Stakeholders' Divergent Values and Power

Individuals, groups, and organizations may have both shared and divergent interests in how a system is designed and deployed. For example, among the many stakeholders in facial recognition technologies are the people who appear in the images, different people potentially now classified as similar enough to members of a group in the images, the workers who labeled the images, the researchers (at varying stages of their careers and in different positions) who developed the different systems, and the researchers at universities or in corporations who adopted the box bounding and face print innovations. The stakeholders also include the people who use the technology, people whose images or faces are recognized correctly or incorrectly, the marginalized populations that are not represented, and government agencies that make decisions based on recognized images and faces. Perhaps the largest, most nebulous stakeholder is society itself as it absorbs the many and varied applications of face recognition and computer vision technologies and the ways they are deployed by macro- and meso-scale organizations.[18]

The power to influence outcomes is not uniform among these stakeholders, and so they are not equally able to advance or advocate for their values.[19] For instance, the political, cultural, and economic environments in which facial recognition development were embedded shaped decisions about what and who to fund; in lab settings, Ph.D. students and professors have different degrees of power to set agendas or challenge the status quo; and marginalized groups who could address whether particular projects using image recognition are appropriate—using faces to identify sexuality or gender, for example—are likely not present and empowered to question and convince researchers to drop such research.

More generally, the design of technologies reflects values of the people and institutions that created them. Researchers' relationships with each other shapes their collaborations and scholarly production. Publishing, intellectual property, and reputational norms impact knowledge sharing about computing research. Differences in power among these groups can thus impact the computing research and technology that result.

---

[18] K. Levy, presentation to the committee on March 11, 2021, Cornell University; V. Eubanks, 2018, *Automating Inequality: How High-Tech Tools Profile, Police, and Punish the Poor*, St. Martin's Press, New York.

[19] S. Costanza-Chock, 2020, *Design Justice: Community-Led Practices to Build the Worlds We Need*, MIT Press, Cambridge, MA.

Values Implicit in Technology Design

An important argument from social studies of technologies is that all sociotechnical systems have particular values built into them that condition the possibilities of use.[20] There is nothing intrinsically wrong—or intrinsically right—about technologies reflecting particular values or encouraging particular uses. For instance, the Internet Relay Chat (IRC) and XMPP distributed protocols reflect different sets of values. IRC's underlying distributed algorithms optimize for real time interaction while sacrificing security whereas the XMPP protocol is open and more secure. Because of these design choices, IRC is often used for applications that might be used by people geographically dispersed who have poor network connectivity and wish to share files. This enables a very particular set of communication and social practices. XMPP might be better for organizations with good network connections that require security for private conversations.

Interactivity of the Social Scales and Technology Design

Sociotechnical systems must be considered from many vantage points to account for the interplay of people and the macro, meso, and micro levels of social phenomena that shape the meaning of technologies when they are put into use. Technology design cannot fully be decoupled from the phenomena of social contexts.

For example, the social phenomena of gender identity—people's everyday experience of gender—is constantly evolving with the macro level of legal requirements, the meso level of cultural norms, and the micro level of self-esteem. Yet, many applications of facial recognition technologies rely on classifying gender as a stable, binary category. As a result of their need to classify and causally discriminate according to assumptions about the ground truth of gender,[21] such technologies cannot inclusively reflect the social phenomena of gender identities. These applications thus raise potential problems of bias.

Similarly, attempts to mathematically define and measure bias cannot encapsulate people's full experience of discrimination, because those everyday social experiences are complex and unique to a specific moment and place in time. Social phenomena, like xenophobia, ableism, racism, patriarchy, and cis-/heteronormativity have multiple interwoven, interacting, and overlapping effects. Cisnormativity, for example, helps explain how some expressions of gender can be marginalized by those considered "normal" in a social setting. However, what counts as normal or typical differs across cultures and

---

[20] L. Winner, 1980, "Do Artifacts Have Politics?" *Daedalus* 109(1):121-136, http://www.jstor.org/stable/20024652.

[21] V. Mays, presentation to the committee on May 6, 2021, University of California, Los Angeles.

points in history[22] and the meanings and impact of these forms of gender discrimination change by location and time.[23]

Societal impacts are not aggregated individual impacts. The impacts of sociotechnical systems are not reducible to effects on individuals. The risks and benefits of computing research may have effects at all three levels of social phenomena. The meso level can be a point of significant impact. For instance, workplace guidelines may determine whether some workers must use facial recognition software to log into a secure worksite. Furthermore, computing research and the systems that follow from them may have social consequences even when their impact occurs through absence; for example, through lack of an accessible, secure, and affordable technical option in some community or from an algorithmic model that fails to recognize one's skin tone or face shape.[24] For example, opponents of facial recognition might use privacy, an individual value, to argue against its use. Privacy also has social effects, taking on very different meanings in societies that see a division between public and private life as important to self-development and social connection.[25]

Social-Historical Influences

Sociotechnical systems arise from the particular social and political contexts of their creators and users, historically bound to them. Such systems conform to laws and regulations, and rely on materials, norms, and social conventions that are historically configured.[26] Even the most innovative new research projects build on existing technology and are affected by the decisions that shaped that technology.[27] Social contexts create opportunities (large swaths of images) and limitations (the images are bound to the context of the services) for computing researchers.[28] For example, when researchers went to social media to find images of people to develop faceprints, they inherited the histories of the designs of those systems. They also inherited the evolving demographics

---

[22] J. Butler, 1990, *Gender Trouble: Feminism and the Subversion of Identity*, Routledge, New York; S. Stryker, 2017, *Transgender History: The Roots of Today's Revolution*, Seal Press, New York.

[23] E. Newton, 2015, *Cherry Grove, Fire Island Sixty Years in America's First Gay and Lesbian Town*, Duke University Press, NC; M.L. Gray, 2009, *Out in the Country Youth, Media, and Queer Visibility in Rural America*, New York University Press, New York.

[24] J. Buolamwini and T. Gebru, 2018, "Gender Shades: Intersectional Accuracy Disparities in Commercial Gender Classification," Proceedings of the 1st Conference on Fairness, Accountability and Transparency in *Proceedings of Machine Learning Research* 81:77-91, https://proceedings.mlr.press/v81/buolamwini18a.html.

[25] S. Duguay, 2022, *Personal Not Private: Queer Women, Sexuality, and Identity Modulation on Digital Platforms*, Princeton University Press, NJ.

[26] L. Winner, 1980, "Do Artifacts Have Politics?" *Daedalus* 109(1):121-136, http://www.jstor.org/stable/20024652.

[27] F.W. Geels, 2005, "The Dynamics of Transitions in Socio-Technical Systems: A Multi-Level Analysis of the Transition Pathway from Horse-Drawn Carriages to Automobiles (1860-1930)," *Technology Analysis and Strategic Management* 17(4):445-476, https://doi.org/10.1080/09537320500357319.

[28] D. Forsythe and D.J. Hess, 2001, *Studying Those Who Study Us: An Anthropologist in the World of Artificial Intelligence*, Stanford University Press, CA.

of user generated content, which in turn depend on differences in access to computing technology across society, thus affecting the quality and biases of the resulting models. Data sets built from user-generated content posted to social media inherit the legacies of regulatory regimes (or lack of regulation) as well as the commercial goals of social media corporations.[29] Having been shielded from certain liabilities for hosting user-generated content, the social media companies adopted algorithms and content moderation policies that incentivized particular kinds of image sharing. These algorithms in turn shaped the content and user interactions among those engaging those social media systems.

Time and shifts in power among stakeholders can change any sociotechnical system. For example, Illinois's passage of the Biometric Information Privacy Act in October 2008 added new constraints on private enterprise's collection, use, and sharing of biometric data from people in the state without consent. A class action lawsuit filed in 2015, representing more than 1 million Facebook users in Illinois, resulted in a $650 million settlement for the company's practice of tagging people in photos using facial recognition without users' consent; in other jurisdictions regulators have not agreed to restrict the private collection of biometric data, so people elsewhere have no such recourse.

Continually Produced and Evolved Interactions

Sociotechnical systems exhibit a continuous feedback loop of interactions between the social causes and the effects of technological change. This loop creates a challenge for responsible computing research: social phenomena and their relationships with technologies can seem stochastic and hard to interpret. From the viewpoint of any individual person and any specific technical system, it can seem impossible to predict, let alone prevent, what happens with technologies as they unfold. As a result, attention needs to be paid at the earliest stages of research to all three scales of social phenomena, including the values that they support or hinder. Such attention and a plan for ongoing monitoring and reevaluation by those deploying technologies or otherwise responsible for their governance is needed as research insights make their way into deployed systems and expectations and concerns shift over time. In the case of facial recognition, vendors have found additional uses, for example, in community-led neighborhood watches and other public safety activities. The concerns at the beginning of a technology's developmental lifecycle are not the same as the ones that surface after wide-scale deployment. As laws, cultural norms, personal experiences, and other social phenomena react to and absorb technologies as they land in people's everyday lives, the uses of any given technology are

---

[29] T. Gillespie, 2018, *Custodians of the Internet: Platforms, Content Moderation, and the Hidden Decisions That Shape Social Media*, Yale University Press, New Haven, CT.

likely to change, in a constantly shifting and evolving ecology that entwines technologies, people, and social contexts.[30]

Predictability of Future Use

A technology's design shapes possible uses, but its interplay with social phenomena is a key factor in determining actual uses. For example, Wikipedia was enabled by wiki technology, but its success was not ensured by that alone. Such micro phenomena as personal interests in sharing facts about Anime and meso phenomena as educated people with leisure time and the organized working structure of Wikipedia were critical.[31] As a result, it is not possible to predict the full set of future uses of any computing research result based on the technical results alone, as social changes could open new possibilities. As a consequence, systems of accountability, equipped to track computing research as it iterates, are critical to addressing its social impact.

All research products have characteristics and capabilities that privilege certain uses in their designs. Particular social, legal, cultural, economic, and political conditions are required for technologies to work in the ways their designers envisioned. Technologies can also be used in ways that researchers, designers, and builders do not fully expect. A sociotechnical approach and the methods and analytic tools of the social sciences enable hypothesizing ways that technology might be used and identifying uses likely to align with value choices and salient macro, meso, and micro level social phenomena.

* * *

The concepts and methods of ethical and sociotechnical analyses presented in this chapter complement one another as essential constituents of responsible computing research in this era of widely deployed and highly networked computing systems. The sociotechnical perspective described in Section 2.2 along with ethical analyses of values and trade-offs as described in Section 2.1, combined with methods of ethical reasoning and such social science methods as ethnographic observation, in-depth interviews, survey studies, and historical analysis, can support computing researchers in identifying and resolving the ethical and societal impact challenges that arise from introducing novel technologies into social worlds. Chapter 3 illustrates their use in identifying underlying roots of such challenges.

---

[30] T. Hughes, 1989, "The Evolution of Large Technological Systems," Pp. 51-82 in *The Social Construction of Technological Systems: New Directions in the Sociology and History of Technology* (W.E. Bijker, T.P. Hughes, and T.J. Pinch, eds.), MIT Press, Cambridge, MA; M. Finn, 2018, *Documenting Aftermath: Information Infrastructures in the Wake of Disasters,* MIT Press, Cambridge, MA.

[31] B.M. Hill, 2013, "Essays on Volunteer Mobilization in Peer Production," Ph.D. dissertation, Massachusetts Institute of Technology.

Together, these concepts and methods enable the development of pragmatic practices that can guide researchers in ways to carry out socially attuned computing research. It is important to note again that computer scientists cannot be expected to become expert ethicists and social scientists. Rather, responsible computing research requires that they collaborate with experts in other disciplines who can bring these important instruments to bear as computing research is designed and carried out.

3

Sources of Ethical Challenges and Societal Concerns for Computing Research

This chapter aims to illuminate the underlying causes of ethical and societal challenges for responsible computing research, grounding them in the ethical and sociotechnical concepts and analyses of Chapter 2. An understanding of these roots is essential to identifying practical steps that those who support and conduct this research can take toward addressing those concerns. The chapter's discussion also makes evident the importance for computing research of incorporating into it consideration of ethical values and trade-offs, of methods from the social sciences, as described in Chapter 2, and of the multidisciplinary collaborations important to realizing ethically and societally responsible technologies and avoiding potential negative consequences of novel technologies. The chapter thus lays a basis for appreciating the report's recommendations.

As noted in Chapter 1, the multi-step translation of research results into deployed algorithms, devices, and systems is effected by researchers, research sponsors, entrepreneurs, investors, and corporate leaders. Actions by participants at any stage of the translation can affect the ethical and societal impact characteristics of any system that emerges from this translation process. The participants in the development and deployment of new technologies, whether individual people or corporations, draw not only on foundational science and engineering research, but also on relevant governmental and corporate governance policies and on legal regulations in determining the shape of a technology. As a result, actions they take depend not only on computing science and engineering but also on those societal level policies and regulations. Furthermore, an additional group of actors plays a role in the deployment of systems: the purchasers of those systems. Ethical and societal impacts are determined by technology choices individuals, organizations, and governments make and the ways in which they use the technologies and systems they acquire.

This chapter identifies a variety of situations, conditions, and computing practices that have potential to raise ethical and societal impact concerns and indicates responsibilities computing researchers have in addressing them. In many cases, the discussions identify challenges that arise because of decisions by those involved in downstream product design, deployment, and acquisition, many of them identified in presentations to the committee (see Appendix B). Some might object that computing researchers have no roles or responsibilities when it comes to downstream product design, implementation, or deployment. Even in these cases, however, computing researchers have obligations. Although they cannot prevent all development and deployment problems, they can minimize the likelihood of misinterpretation or misuse of their research by others through clearly delineating the limitations of the capabilities and the intended scope of applicability and use of the methods and artifacts their research yields. Their in-depth knowledge of their research also places them in a unique position to inform the public and advise government on such facets of these methods and artifacts as their intended situations of use, scope of validation, and limitations.

Computing research itself is embedded in a range of social contexts: the university or research organization in which it is being carried out, the organization funding the research, and the society in which these organizations exist form an ever evolving and complex human system. These contexts yield a range of societal factors that affect the ethicality and societal impact of the research across the four groups. They also influence choices of research problems, the membership and structure of the teams that conduct the research, and the stakeholders engaged during the research process.

The chapter divides ethical and societal impact challenges into four groups: those that arise from (1) features of the societal settings into which new computing technologies are introduced; (2) limitations of human capabilities and ways they interact with features of computing technologies and the societal contexts in which computing technologies may be used; (3) features of the societal contexts into which computing technologies may be integrated that influence computing system design and deployment choices; and (4) system robustness problems caused by failure to follow best practices in design and implementation.

The final section of the chapter highlights the limits of purely computing-technical approaches to meeting the challenges of societally responsible computing technologies. It provides two examples, each of which illustrates the need for policy and regulation to work in tandem with computing technology design and deployment.

3.1 SOCIETAL CONTEXTS

The social ecosystems in which computing technologies participate give rise to challenges rooted in the diversity of human participants in these sociotechnical systems, which yields the possibilities for conflicting values and goals, the need to respect human dignity, and complexities for predicting individual behavior; a recognition that although computing technologies can help address societal challenges, they can do so only to a certain extent; the influences of the institutional structures and norms into which computing technologies may be integrated; and the societal-level impact these technologies might have.

The examples given in this section make apparent a range of ethical and sociotechnical factors to which computing researchers should pay attention, including the need for inclusiveness in research of the various stakeholders potentially affected by research outcomes, the importance of elucidating limitations of research results as well as contributions and benefits, the need for computing researchers to consider the potential extreme societal level harms of computing, the importance of reshaping education and training in computer science and engineering, and the resulting needs to include multidisciplinary expertise in computing research, reshape computing education and training, and to assist the public's and government's understanding of research outcomes. In doing so, it both reaches back to the foundations provided in Chapter 2 with examples of the general principles that chapter lays out, and looks forward to the recommendations in Chapter 4.

3.1.1 Reconciling Conflicting Values and Goals of Stakeholders

As noted in Chapter 2, in a pluralistic society, different individuals or groups may have very different values and interests. In any context, some values and interests can justifiably take priority over others. For example, in everyday life, one person's interest in privacy almost always supersedes another's interest in idle curiosity. All technologies, including computing technologies, prioritize particular values and interests. In her remarks to the committee, Sarah Brayne observed that, before turning to questions about technology, one must answer normative questions. In criminal justice, is the goal of deploying computing technology to reduce crime, reduce prison populations, better allocate scarce policing resources, or something else?[1] Ece Kamar also pointed to a workplace surveillance technology conflict; such technologies can be used to incentivize only productivity gains or to also collect information useful for coaching employees in ways that foster their professional development.[2]

---

[1] S. Brayne, presentation to the committee on March 4, 2021, The University of Texas at Austin.
[2] E. Kamar, presentation to the committee on March 11, 2021, Microsoft Research.

Computing researchers make value choices even when their research is not explicitly aimed at an application. A research project focused on faster chips prioritizes computing power; a research project focused on chips that use more environmentally friendly materials prioritizes environmental benefits. Furthermore, value conflicts can be resolved in many different ways, and as subsequent sections discuss, a variety of stakeholders' interests should play a role in decisions about which values to prioritize in computing research.

As Chapter 2 argues, computing research is itself not value-neutral, and also computing technologies have many different kinds of stakeholders. Computing researchers and the computing research community prioritize, often implicitly, some values over others, with such choices occurring not only in designing systems or empirical investigations and data gathering, but also in the choice of research problem area and the particular aspect of the problem considered. The more diverse the group involved and empowered in these choices—disciplinarily, demographically, and geographically—the more likely the group is to notice mistaken assumptions and recognize biases, and thereby act on them.

Last, it is important to recognize that it is not an option to somehow prioritize everyone's values equally. Computing technologies and research necessarily prioritize certain values over others. As an analogy, consider building a restaurant menu. The placement of items on the menu influences people's orders, so one could build a menu that makes healthy choices more likely, or more profitable choices more likely, or advances some other values in terms of its influence on choices. But one cannot choose to prioritize all values equally (unless they all happen to coincide), because one has to pick *some* layout for the menu items. Similarly, one cannot conduct value-free or "value-uniform" computing research, and so cannot avoid deciding which values and interests will, for the purposes of this research, be prioritized.

3.1.2 Preserving Human Dignity

The concept of human dignity is rooted in human intuitions and deeply held values as well as domestic and international law. Human rights documents such as the Universal Declaration of Human Rights underscore that concepts of dignity often transcend cultural and national differences, even if those differences also affect nuances in how particular aspects of the concept are understood or implemented. Dignity plays an important role in the American legal system, reflected in discussions of American legal commitments to due process, for example, and protection against unreasonable searches and seizures. Concerns about human dignity also arise well beyond the legal system, in domains as diverse as civic education, research on ethics, and broad efforts to reform public institutions and safeguard privacy in civil society. Human dignity is more than merely a subjective feeling. It encompasses a variety of closely related concerns, including the intrinsic ethical values of autonomy, well-being, and justice and legitimate power, as well as such instrumental ethical values as privacy, safety, and security.

Computing technology can affect all these elements of human dignity. Computer systems and the data they use can affect labor and the marketplace, shape social activity, and structure the public's relationship with their government and each other. Changes in applications, systems, and data have the potential to alter fundamental aspects of dignity, including privacy, freedom and agency, and physical security. For example, better design can afford better access for users with different usability needs, such as those with vision, hearing, or mobility impairments (see Section 3.4.4). Computing research can affect the downstream characteristics of these applications, systems, and data. To the extent that choices among values and prioritizations choose one group's preferences over another, they may risk sacrificing some people's human dignity. These issues reflect the importance for computing researchers to consider the full range of potential stakeholders and the diversity of values they may hold for technologies that incorporate the results of their research, as well as the need for multidisciplinary expertise in shaping computing research projects that could potentially affect these values.

A range of examples help illustrate connections between computing research and potential impacts—both positive and negative—on human dignity. On the positive side, adjustments to workplace scheduling software to give more control to workers can significantly improve workers' quality of life and sense of dignity.[3] On the negative side, any strictly rule-based system in the administration of criminal justice will likely be both over- and under-inclusive, and so there need to be humans in the loop to ensure that all parties retain their human dignity.[4]

The double-edged potential of computing research for human dignity is readily seen in possible uses of automation to help handle the crush of civil litigation.[5] On the one hand, many people currently lack realistic access to human expertise in situations where civil litigation might be appropriate for them, and so computing technology could help restore their autonomy and dignity. On the other hand, removing the human element of the civil litigation process might erode social stability and legitimacy, thereby harming everyone's human dignity.

Even the process of research can affect the dignity of individuals whose data or behavior are used, as they are often treated as mere data points, rather than humans with inherent dignity. Computing researchers thus need to carefully consider the potential

---

[3] K. Levy, presentation to the committee on March 11, 2021, Cornell University.
[4] A. Roth, presentation to the committee on March 4, 2021, University of California, Berkeley, School of Law.
[5] B. Barton and G. Hadfield, presentation to the committee on May 11, 2021, University of Tennessee and University of Toronto.

consequences of their research projects—both positive and negative—for human dignity.

3.1.3 Responsibly Predicting and Shaping Individual Behavior

Data-intensive computing methods, including machine learning, enable predictions that have the potential to be more accurate and less shaped by cognitive biases and heuristics, as well as by explicit or implicit discriminatory attitudes. Systems deploying these methods do not commit the base rate fallacy or adopt the availability heuristic.[6] They have no intrinsic racist animus.[7] They can detect patterns in massive data sets that are impossible to identify using other methods. With the vast expansion in computing power, the availability of large amounts of data, and the availability of predictive models, such methods seem easy to deploy in new areas and are being used to improve predictions of many sorts for which training data is available. The absence of human discriminatory animus explicitly in these systems, however, in no way ensures that those systems will not reproduce, in digital form, structurally biased social phenomena, as critiques of such technologies as facial recognition and predictive policing show.[8] Computing research avenues for addressing this challenge include more intentional participation of diverse users in training systems and development of objectives for model training that counter such biases.

Predictions of large-scale social phenomena are vital contributors to public policy and are central to social scientific research. Advances in computing systems, however, now offer the promise of predicting the behavior not of populations, but of individuals. And these acutely individualized predictions have been mobilized in contexts with very high stakes (ones that put at risk many of the ethical values in Section 2.1)—such as decisions over whether to release a defendant pretrial, whether to admit a student to the university, and whether to award or deny an application for credit. Computing researchers working to predict the behavior of individuals must be especially cautious about how their research will be used.

At the scale of public policy, the impact of predictive models on the population as a whole can be assessed, with errors being straightforwardly compensated for by improvements over other predictive tools. At the scale of decision-making about individuals' lives, a mistaken prediction cannot be likewise compensated for. This difference is one reason for long-standing practices of due process in the legal system that go far beyond merely applying globally optimal probabilistic tools. At the population level, the performance of the model over time can be validated through established statistical

---

[6] D. Kahneman, P. Slovic, and A. Tversky, 1982, *Judgement Under Uncertainty,* Cambridge University Press, MA.
[7] J. Kleinberg, J. Ludwig, S. Mullainathan, and C.R. Sunstein, 2019, "Discrimination in the Age of Algorithms," *Journal of Legal Analysis* 10(2018):113-174, https://doi.org/10.1093/jla/laz001.
[8] S.U. Noble, 2018, *Algorithms of Oppression,* New York University Press.

methods. At the individual level, there is no ground truth against which to validate any particular intervention: if a defendant is not released pretrial, it is not possible to determine whether they would have been rearrested, had they been released. This fact does not imply that such models should not be developed, but that when they are developed or deployed, if at all, computing researchers, developers, and deployers should do so with full knowledge of the significant moral risks associated with predicting individual behavior in uses that make high stakes decisions about people's lives; researchers should therefore clearly delineate the limitations of the capabilities and the intended scope of applicability of their methods and systems.

Second, computing systems are also being used to predict human behavior in order to calibrate interventions that are then used to shape that behavior. This can range from supercharging "nudging" to take account of particular characteristics of users,[9] to designing recommender systems aiming to optimize user engagement, or advertising systems focused on increasing click-through rate.[10] Actors in society have always used any means available to shape others' behavior; the concern with advances in computing research, combined with our increasing dependence on computing systems, is that the means now available for this purpose are much more effective, more pervasive, and more readily available. Even if computing systems are relatively ineffective at manipulating any particular individual, they are demonstrably effective at a kind of stochastic manipulation, whereby populations are moved by modestly influencing the behavior of their constituents.[11]

That computing systems may be used to make high stakes, highly sensitive, and highly risky predictions about individuals' lives, and their potential use to calibrate digital interventions intended to shape people's behavior are concerns not only for system development or deployment. If computing researchers become aware that their work could potentially be used to predict human behavior for these purposes, they need to consider the ethical ramifications of such uses and consider the context in which their research is done and the potential implications for society, groups, or individuals, and the values they are trading off. In many cases, it will be vital to build in robust procedural protections for those affected, because there is no ground truth against which to validate any particular intervention.[12] Ensuring that their predictive models of human behavior

---

[9] K. Yeung, 2017, " 'Hypernudge': Big Data as Regulation by Design," *Information, Communication and Society* 20(1):118-136, https://doi.org/10.1080/1369118X.2016.1186713.

[10] T. Graepel, J. Quiñonero Candela, T. Borchert, and R. Herbrich, 2010, "Web-Scale Bayesian Click-Through Rate Prediction for Sponsored Search Advertising in Microsoft's Bing Search Engine," Pp. 13-20 in *Proceedings of the 27th International Conference on Machine Learning*, https://www.microsoft.com/en-us/research/wp-content/uploads/2010/06/AdPredictor-ICML-2010-final.pdf.

[11] C. Benn and S. Lazar, 2021, "What's Wrong with Automated Influence," *Canadian Journal of Philosophy* 1-24, https://philpapers.org/rec/BENWWW-3.

[12] D.K. Citron, 2008, "Technological Due Process," *Washington University Law Review* 85(6):1249-1313, https://journals.library.wustl.edu/lawreview/article/id/6697.

are appropriately auditable as well as interpretable by users is a challenge computing researchers need to address. (See Section 3.2.4, "Understanding Behavior of Opaque Systems.") These issues reflect not only the importance of considering a range of ethical values and trade-offs among them as well as the full range of potential stakeholders, but also the need for multidisciplinary expertise and for researchers to make clear the intended uses and the limitations of their research outcomes.

3.1.4 Proper Roles for Technologies in Addressing Societal Problems

Technologies can help address social problems, but technological approaches cannot solve societal problems on their own, as evident from the discussion of sociotechnical systems (see Section 2.2). There are numerous examples of problems caused by unconstrained technological solutionism—the default tendency to pursue purely technological solutions to societal problems. Ben Green's book *The Smart Enough City*[13] describes how some of these difficulties[14] play out in the setting of urban planning; similar challenges arise in many other settings.[15]

In designing technological interventions, it is important to distinguish between approaches that address the symptoms of societal problems and approaches that address the root causes of these problems. Consider, for example, the rapid growth of legal automation and arbitration software, aiming to address the radical under-provision of court time or affordable legal representation to resolve private disputes in the United States, or the development of legal aid software aimed to simplify complex legal documents for the large number of individuals who cannot afford a lawyer. Computing may be able to help address these social problems, but to do so it needs to be designed to support people in roles that address those problems. Even so, there are deeper underlying problems beyond the sphere of computing research—the litigiousness of a society, the large volume of small-claims disputes, and difficulties interpreting legal documents—that need to be addressed.

Computational approaches rely on abstracting problems to make them formally representable and mathematically tractable. This process of abstraction is both a crucial source of power in the design of solutions to computational problems and a significant source of risk for societal problems, as models can omit details that ultimately prove crucial or unintentionally focus on symptoms rather than underlying causes. Here too there are numerous examples of models that produce disastrous results because crucial

---

[13] B. Green, 2019, *The Smart Enough City: Putting Technology in Its Place to Reclaim Our Urban Future*, MIT Press, Cambridge, MA.

[14] E. Morozov, 2013, *To Save Everything, Click Here: The Folly of Technological Solutionism*, PublicAffairs, New York.

[15] See, e.g., S. Brayne, 2021, *Predict and Surveil: Data, Discretion, and the Future of Policing*, Oxford University Press, United Kingdom.

domain specifics were omitted. To take another example cited in Ben Green's writing, late 19th-century loggers in Germany replaced ancient forests with managed alternatives, mathematically optimized for maximum timber yield per acre. The models on which the optimization was based were irremediably flawed and failed to represent the importance of biodiversity in sustaining forest ecosystems; after two generations the forests died and could not be revived.[16] This example is also a reminder that one must monitor impacts and adjust in accordance with such feedback.

These examples illustrate that for computing research to support the development of computing technologies that address societal problems, it needs to take into account the social contexts in which the models, methods, or artifacts it produces might be deployed. For them to be able to address societal problems adequately and ethically for all potentially affected populations requires that all potential stakeholder populations are included in design decisions (whether of algorithms or of computing systems). The examples thus also show the importance of research that aims to address societal problems being multidisciplinary from the start.

3.1.5 Aligning with Existing Norms, Structures, and Practices

Social relationships may be mediated by various technologies as Chapter 2 describes. When computing science researchers are thinking about developing new systems, they should work with social scientists and stakeholders to understand the relevant existing social norms, institutional structures, and practices. Institutional structures and existing norms can be buttressed or challenged by novel computing technologies. New technologies can alter power structures, shift work loads, change compensation incentives and affect the nature of work. Integration of new technologies done well can improve structures and practices; done poorly it can obstruct them. In many domains, including health care, work and labor, and justice, the introduction of new technologies can affect equity and inclusion.

Examples of computers reifying existing power structures are prevalent in discussions of work and labor. Employees might be surveilled and the information that is gathered is only shared by managers (who are not surveilled). This can give managers knowledge that workers do not have. Karen Levy's remarks to the committee revealed that data-intensive technology in the workplace can be used to transfer burdens that have generally been borne by managers and owners to workers. Levy says that technologies that are ostensibly enacting efficiencies place extra burdens on workers. A bevy of other examples cited by Levy illustrate how social and economic relationships are being reorganized by computing technologies: Amazon used to compensate authors who

---

[16] J.C. Scott, 2008, *Seeing Like a State,* Yale University Press, New Haven, CT.

published with their platform based on how many people downloaded a title, but then switched to only compensate authors for the number of pages of books that get read. (In music, this shift has altered the type of cultural products that are produced to navigate away from albums and navigate more toward individual songs.) Lastly, Levy discussed how computing technology has recently been deployed to monitor the tone of call center workers and how such monitoring could be quite overbearing and create an unproductive and inhuman psychological burden on those workers. Computing technologies shape social relationships between manager and employee in these examples.[17]

One area where computing technologies have not easily integrated with existing institutional structures and norms is health care. Perhaps the most familiar example is electronic health records (EHRs), which have had a slow uptake and required an enormous amount of funding to widely deploy. In-home monitoring is another emerging use of computing technology, which has been propelled in part by the health-conscious patients who have been among the early adopters. Examples include digital watches, digital scales, and digital glucometers. As Robert Wachter explained in his remarks to the committee, doctors increasingly can monitor their patients between visits, making a great deal more information available that can be used to enhance treatment. The technologies allow the reach of health care to be extended from hospitals and medical offices to patients' homes. However, this development raises questions about whether patients and their families will choose this type of health care relationship.[18]

In her remarks to the committee, Abi Pitts described her work during the COVID-19 pandemic with children in foster care and other precarious situations. The move to telemedicine was challenging because vulnerable patients did not always have access to Internet-connected devices, yet these technologies were integral to the main mode of care delivery. Further complicating the challenge of serving patients was the requirements of electronic health record systems and other kinds of medical technologies to have stable and caring parents or guardians as points of contact and to provide the consent required for minors in need of health care. The move to telemedicine resurfaced the need for more flexible approaches to adult oversight than the available technology platforms allowed.[19]

Madeline Clare Elish's presentation to the committee on the SepsisWatch project illustrated just how much careful work and planning is involved in successfully integrating a new technology into a hospital setting. She discussed how a successful deployment of new computing technologies can be contingent on changes to existing social and institutional structures. The effective integration of the tool into hospital workflows

---

[17] K. Levy, presentation to the committee on March 11, 2021, Cornell University.
[18] R. Wachter, presentation to the committee on March 16, 2021, University of California, San Francisco.
[19] M.A. Pitts, presentation to the committee on May 11, 2021, Stanford University School of Medicine/Santa Clara Valley Medical Center.

was not possible until the expert knowledge of nurses was reflected in new practices and procedures.[20]

These observations about computing technology in work and health care settings exemplify the importance for sociotechnical systems design of including all potential stakeholders from the beginning of computing technology design and of multidisciplinary teams. They also illustrate the limitations of computing research's abilities to ensure responsible computing technologies and the resultant need for informed corporate policy and government regulation, to which the expertise of computing researchers can contribute. Recommendation 8 indicates ways in which researchers can contribute to the development of such policies and regulations.

3.1.6 Addressing Environmental Externalities

Computing research has generally not focused on environmental externalities but in recent years, attention has focused on the energy and materials use of all computing systems. To address environmental externalities, computing researchers will need to broaden how they view the impacts of computing technologies and consider the wide range of ethical values and trade-offs raised by these externalities. Significant strides have been made to enhance the energy efficiency of data centers and other infrastructure, drawing on computing research on more energy efficient components, architecture, and power and cooling designs as well as using renewable energy sources. At the same time, the energy consumption of this infrastructure continues to expand with growing demand for computing and storage. Battery constraints have propelled significant improvements in the power performance of laptops, smartphones, and embedded devices as well as the power performance of specific algorithms, architectures, and system designs. Yet, not all computing research, even research involving systems such as Bitcoin that are known to be extremely power-hungry, necessarily takes energy externalities into account.

Computers require a wide array of materials, some of which are extracted only at great cost in energy or come from mines and processing facilities that may pollute, provide poor working conditions, employ child labor, or fuel conflict. Commercial technology producers have few incentives to design and sell objects that are made to last, are recyclable, or can easily be repaired. The result is even more demand for materials as well as pollution from discarded items. Right to repair movements have blossomed from critiques of this mode of commerce, and some computing researchers have suggested new paradigms for technology designs that move away from "disposable" technology and enable longer life cycles, repair, and use sustainable materials. Researchers in fields

---

[20] M.C. Elish, presentation to the committee on March 16, 2021, Google, Inc.

such as human–computer interaction (HCI) have also challenged researchers to consider the use of materials and opportunities for reuse and recycling in processes such as prototyping and design.

Computing in a world of limited resources also raises questions about what computing capacity is necessary or socially desirable. Should every user's click on a random website be saved for decades? Are the benefits of power-hungry, proof-of-work blockchain systems such as Bitcoin worth the environmental price? Is it worth making devices harder to repair in order to make them thinner or lighter?

Although a source of the unsustainable consumption that drives climate change, computing is also a key tool for understanding and managing that change. Climate analysis relies on enormous data sets and complex computational models, and computing is also an important tool for optimizing and managing energy-consuming activities in order to reduce their carbon footprint.

3.1.7 Avoiding Computing-Related Extreme Events

Computing technologies can be associated with scenarios involving risks of extreme events—destructive failures and similar outcomes that are massively costly for society. At present, these events are likely to arise as a result of misuse (intended or not), failure, or unexpected properties that emerge from large, complex networks. Examples familiar from recent headlines include ransomware attacks that shut down courts or hospitals for days or weeks and computer failures causing the electric grid to fail for an extended period and flash crashes of stock markets. An example of a potential extreme risk would be a computing malfunction or cyberattack affecting the command and control of nuclear weapons.

These examples are a reminder that computing researchers should not dismiss the possibility of catastrophic harms that may result from the imperfections or emergent and unanticipated properties of sociotechnical systems. By educating government policymakers, and monitoring uses of computing technologies, computing researchers can help society better manage these risks. Although current computing researchers are not responsible for continued use of obsolete computing technology (e.g., in air traffic control or nuclear weapons command and control), they can contribute to the much-needed better development of good risk assessment tools going forward. Given that the risks depend on the whole sociotechnical system (humans and computers), further research and analysis will bring greater precision to the assessment of risks involving computing technologies.

There is also some possibility that certain extreme risks will arise not only from deployment of computing innovations, but from the research process itself. While in the end the Morris worm[21] was not so destructive, its origin as a research project is a reminder that research may inadvertently cause harm, a possibility that the continuous integration of research results into deployed systems makes more likely.

3.1.8 Influences of Social Structures and Computing Pedagogy

This section examines two organizational facets of computing research of which responsible computing research requires computing researchers and the computing research community to be aware.

Advancing Diversity, Equity, and Inclusion

For computing research to be responsible requires taking into account the influence of a variety of organizational and social structures on decisions made in carrying it out. Two such structural influences on computing research in particular raise significant ethical and societal impact issues: the lack of diversity of the computing research community and the lack of inclusiveness of affected populations in the design, development, and testing of computing research and the artifacts it produces. Insufficient awareness of and lack of attention to systemic racism and sexism—which includes the implicit perpetuating of biases that were once explicit and blatant—have raised and continue to raise ethical and societal impact concerns.[22]

Computing researchers and the computing research community at large cannot by themselves change these structures. Redress requires action at every level from society and government to individual researchers, and much collaborative work. The discussion in this section highlights issues that the computing research community needs to be aware of so it can be mindful of ways it can adjust processes of research, system design, and deployment to mitigate harms.

Diversity has many different dimensions, including race, gender, ethnicity, geography (e.g., the Global South is often excluded from decisions about computing technologies), and cognitive and physical capabilities. Although there may remain questions about the extent to which various types of diversity contribute to responsible computing research, diversity is essential to better problem solving and design.[23] The negative

---

[21] Federal Bureau of Investigation, "Morris Worm," *Federal Bureau of Investigation,* https://www.fbi.gov/history/famous-cases/morris-worm.

[22] See, for example, R. Benjamin, 2020, *Race After Technology Abolitionist Tools for the New Jim Code,* Polity, Cambridge, United Kingdom; W. Asprey, 2016, *Women and Underrepresented Minorities in Computing: A Social and Historical Study,* Springer, New York; J. Abbate, 2012, *Recoding Gender,* MIT Press, Cambridge, MA; T. Misa, ed., 2020, *Gender Code: Why Women Are Leaving Computing,* Wiley-IEEE Press, Hoboken, NJ.

[23] See, for example, D. Rock and H. Grant, 2016, "Why Diverse Teams Are Smarter," *Harvard Business Review* 4(4):2-5, https://hbr.org/2016/11/why-diverse-teams-are-smarter; A.W. Woolley, C.F. Chabris, A. Pentland, N. Hashmi, and T.W. Malone, 2010, "Evidence for a Collective Intelligence Factor in the Performance of Human Groups," *Science* 330(6004):686-688, https://doi.org/10.1126/science.1193147.

impacts of lack of inclusiveness are also clear now that broad swaths of that research and its use increasingly involve the collection, computational analysis of that data, and its use. The biases that have been found in various face recognition and judicial sentencing systems are reminders of the problems that lack of inclusiveness in research design and lack of diversity on research teams can cause.

Computing research remains a field dominated by White men, despite various efforts—including the National Science Foundation's significant investments in its Broadening Participation in Computing program—to diversify. Small, token representations of currently underrepresented groups do not work. Until people from a broader range of perspectives and backgrounds make up a significant portion of a group, their ideas and perspectives are likely to be frequently overlooked. This state of affairs affects both the pipeline and the challenge of retaining talent. The resulting computing research environment influences the perspectives that are brought to bear on any project, including what are understood to be the "best" or "appropriate" computing research priorities.

The related problem of inclusion in relevant aspects of computing research design, implementation, and deployment reflects the fact that increased diversity on a research team by itself is not sufficient. Differences in race and gender are not simple differences, but structurally hierarchical ones that almost always yield power and importance differences. A lack of attention to structural racism and sexism have led to many current problems of computing technology, problems which often have origins in a similar lack of attention in the computing research that underlies those systems. For instance, a health risk assessment model based on hospital admission data that was used as a predictor of who would need a hospital bed was developed using a data set that disproportionately favored people with insurance. This data set bias resulted from overlooking the fact that race plays a big role in who has insurance and can therefore afford to be hospitalized, and thus who will be admitted. As a result of overlooking this structural bias in the data, in its predictions, the model perpetuated an existing bias. In algorithmic screening of job candidates, it has proven extremely difficult to remove bias.[24]

Similarly, the design and use of predictive algorithms used in pretrial release determinations or sentencing assessments must take into account structural factors that affect both the data used to train their models and how those tools are used. For instance, because one is more likely today to be arrested or stopped without cause if Black than if White, feeding such data into models used for future predictions about criminality will replicate racial bias.[25]

---

[24] M. Raghavan and S. Barocas, 2019, "Challenges for Mitigating Bias in Algorithmic Hiring," The Brookings Institution, https://www.brookings.edu/research/challenges-for-mitigating-bias-in-algorithmic-hiring.

[25] R. Hutchins, presentation to the committee on March 4, 2021, University of the District of Columbia School of Law.

Inclusion, ensuring that working environments foster participation by and advancement of members of underrepresented and historically marginalized groups is also important in design. For instance, it requires that computing research as well as technology development recognize that social inequities yield differences in access to technology from an array of sources including lack of connectivity, app interface inaccessibilities (e.g., for those with vision or hearing challenges), and language and dialect variations that are not adequately addressed in language and speech processing technologies. Absent inclusiveness of people whose abilities, perspectives, and experiences are different from researchers, results are likely to perpetuate or exacerbate existing inequities. By contrast, efforts to counter structural ableism in widely used computing platforms have led to much improved HCI for all users.[26]

Diversity, equity, and inclusion issues arise at all three sociotechnical levels described in Section 2.2. Long-standing cultural prejudices typically underlie lack of diversity and inclusion, and market and scholarly incentives may influence choices researchers make. Although structural responsibility does not preclude individual responsibility, social structures are often organized in such a way that another person in a similar position would make the same ethically problematic choice. The social facts that engender organizational problems are typically generated over time, often through uncoordinated acts of many individuals and groups. Some responsibility for the problems thus rests in these social structures of the organizations within which researchers carry out their work. To bring about different outcomes requires organizational commitments to changing research environments.

Integrating Ethical and Societal Issues into Training

Computing researchers are commonly trained in computer science, computer engineering, or a related discipline. Typical curricula and research training in computer science and engineering still involve little or no exposure to the social and behavioral sciences. The absence of significant exposure to ethical and societal impact issues and to methods for identifying and addressing them is one possible explanation for computing researchers neglecting them in their research. Furthermore, and not surprisingly, most computer science researchers consider themselves unqualified to teach ethical and societal implications of their work, and many consider doing so to be outside of their sphere of responsibility.[27] Some institutions have responded by adding an introductory course in technol-

---

[26] This opportunity was realized several decades ago. See National Research Council, 1997, *More Than Screen Deep: Toward Every-Citizen Interfaces to the Nation's Information Infrastructure,* National Academy Press, Washington, DC, https://doi.org/10.17226/5780, p. 41.

[27] C. Ashurst, E. Hine, P. Sedille, and A. Carlier, 2021, "AI Ethics Statements—Analysis and Lessons Learnt from NeurIPS Broader Impact Statements," arXiv, https://doi.org/10.48550/arXiv.2111.01705; M. Hutson, 2021, "Who Should Stop Unethical A.I.?" *The New Yorker,* https://www.newyorker.com/tech/annals-of-technology/who-should-stop-unethical-ai.

ogy and ethics. Such courses may be useful, but as is discussed below, there are both practical and pedagogical disadvantages to relying on a single course that is divorced from the core courses in computer science and engineering. If optional, the students who need it most may not take it; if required, it may be difficult to fit into an already crowded curriculum.

A substantial literature on computing ethics has developed since Moor's 1985 paper[28] argued for it as an emerging important area of philosophical scholarship. Although computing ethics is an independent area of research in its own right,[29,30] there is an ongoing need for more practically engaged and technically informed philosophical scholarship to help computing researchers better understand the ethical implications of their work.

For responsible computing research, there is an urgent need for computing researchers to acknowledge that considering the societal and ethical implications of their work is an essential component of that work. This change of mindset, however, is only a first step. To identify and address ethical and societal impact issues, computing researchers need their research projects to incorporate relevant methods and tools from a broader range of disciplines. Research organizations need to make structural changes in the ways they organize research efforts, including providing incentives for scholars and researchers in all relevant fields to engage in such efforts, including modifying the way researchers are evaluated for promotion and tenure.

The ways in which to effect such multidisciplinary efforts depend on career stage. Established researchers cannot be expected to become expert in other fields, but they can learn and benefit from acquiring sufficient knowledge of the approaches of those fields to identify those with whom they should collaborate on a given project. Recommendations 3.2 and 3.3 in Chapter 4 include several possibilities for them to acquire competencies in working with scholars in the humanities and social sciences.

Students (undergraduate, graduate, and postdoctoral trainees) need broader educations than the heretofore standard ones in computer science and engineering. Neither the need to broaden the computer science curriculum to encompass ethics and consideration of societal impact nor the challenges to doing so are new,[31] but the need has become more urgent. The burgeoning interest in ethics and societal impact among students makes this time a propitious one for introducing such changes to the

---

[28] J.H. Moor, 1985, "What Is Computer Ethics?" *Metaphilosophy* 16(4):266-275, https://onlinelibrary.wiley.com/doi/abs/10.1111/j.1467-9973.1985.tb00173.x.

[29] B.C. Stahl, J. Timmermans, and B.D. Mittelstadt, 2016, "The Ethics of Computing: A Survey of the Computing-Oriented Literature," *ACM Computing Surveys* 48(4):1-38, https://dl.acm.org/doi/pdf/10.1145/2871196.

[30] R.T. Pennock and M. O'Rourke, 2017, "Developing a Scientific Virtue-Based Approach to Science Ethics Training," *Science Engineering Ethics*, https://link.springer.com/content/pdf/10.1007/s11948-016-9757-2.pdf.

[31] K. Miller, 1988, "Integrating Computer Ethics into the Computer Science Curriculum," *Computer Science Education* 1(1):37-52, https://doi.org/10.1080/0899340880010104.

curriculum. Internships along with course work can help students learn the importance of disciplinary diversity among the people taking part in the computing design process. Computing researchers, working in collaboration with colleagues in the social and behavioral sciences as well as ethics, need to apply the same creativity and discipline to this part of the work as to design, development and analysis. As Eden Medina said in her remarks to the committee, "Indeed, the kinds of quandaries that we see today often cannot be simplified in terms of right or wrong, which is how students sometimes view ethics. In fact, what we often see are people who are acting with good intentions, or who believe deeply in their work, but they don't understand the full implications of their work and from one perspective, how could they possibly, especially since so much about computing is framed as new."[32]

Furthermore, if it is to have the desired effects, the teaching of ethical and societal impact implications of computing research and technology development needs to be integrated into the computing courses across the curriculum and not offered only as "one-off" siloed separate classes.[33] Various projects funded by the Responsible Computing Science Challenge,[34] which are multidisciplinary efforts, have pioneered a variety of promising approaches.[35] The academic institutions funded by this challenge are both public and private, and they are diverse in size, student populations, and resources. Their approaches and others' address a number of challenges to incorporating ethics into computer science and engineering curricula, including scaling across the curriculum[36] and providing the requisite expertise in relevant ethics and social science disciplines without expecting computer science and engineering faculty to develop it. These projects have also identified a variety of ways of reducing barriers and incentivizing participation.

The embedding of ethics into existing courses has four important features:

1. It directly ties ethical and societal impact thinking to course material, so students learn that these considerations are part of computing research and technology development work and so students experience embodying ethical

---

[32] E. Medina, presentation to the committee on May 25, 2021, Massachusetts Institute of Technology.

[33] B. Grosz, D.G. Grant, K. Vredenburgh, J. Behrends, L. Hu, A. Simmons, and J. Waldo, 2019, "Embedded EthiCS: Integrating Ethics Across CS Education," *Communications of the ACM* 62(8):54-61, https://doi.org/10.1145/3330794.

[34] Mozilla, "Responsible Computer Science Challenge," https://foundation.mozilla.org/en/what-we-fund/awards/responsible-computer-science-challenge.

[35] C. Fiesler, 2018, "What Our Tech Ethics Crisis Says About the State of Computer Science Education," *How We Get to Next,* https://www.howwegettonext.com/what-our-tech-ethics-crisis-says-about-the-state-of-computer-science-education.

[36] B.R. Shapiro, E. Lovegall, A. Meng, J. Borenstein, and E. Zegura, 2021, "Using Role-Play to Scale the Integration of Ethics Across the Computer Science Curriculum," Pp. 1034-1040 in *Proceedings of the 52nd ACM Technical Symposium on Computer Science Education*; L. Cohen, H. Precel, H. Triedman, and K. Fisler, 2021, "A New Model for Weaving Responsible Computing into Courses Across the CS Curriculum," Pp. 858-864 in *Proceedings of the 52nd ACM Technical Symposium on Computer Science Education.*

thinking in the various seemingly mundane decisions of their computing practices, so these skills are learned in the same integrated way as computing technical skills;[37]

2. Students who are not computing majors but take computer science and engineering courses and may go on to careers for which understanding of computing and computing research and its ethical implications are important (e.g., in government or in corporate management) also acquire the requisite knowledge and reasoning skills;

3. The integration and distribution throughout the curriculum address failure points identified in stand-alone ethics courses in engineering, including having greater promise of changing the culture than stand-alone courses; and

4. It does not require adding course requirements to computer science and engineering majors, which are often already overfilled.

3.2 LIMITATIONS OF HUMAN CAPABILITIES

Various characteristics of the physical and social worlds in which computing systems operate have potential incompatibilities with certain limitations of human capabilities for designing and understanding these systems. This section describes four arenas in which interactions of people with computing systems raise significant ethical and societal impact concerns. The first focuses on the kinds of situations in which computing systems now often operate, the other three on aspects of human capabilities for interacting or working with computing systems. The discussion and examples presented in the subsections show that responsible computing research requires computing researchers to be transparent about the intended use situations for the computing methods and artifacts their research produces, limitations in their power and applicability, the assumptions about people's capabilities their performance rests on, and the range of situations in which they have been tested. These responsibilities are reflected in Recommendation 6 (especially 6.5) and Recommendation 7 (most notably 7.2).

3.2.1 Designing for Open Worlds

Computing systems are increasingly situated to operate in the physical and social worlds, with all of their complexities and interactions. Invariably, computing researchers will have limited knowledge about the situations in which a system will operate, because

---

[37] L. Bezuidenhout and E. Ratti, 2021, "What Does It Mean to Embed Ethics in Data Science?: An Integrative Approach Based on Microethics and Virtues," *AI & Society* 36(3):939-953, https://doi.org/10.1007/s00146-020-01112-w.

computing systems are now typically used in "open worlds" rather than closed, well-defined environments with a limited set of (usually trained) users. As a result, there is always the possibility of the system encountering an unexpected situation or some additional information that might affect its behavior, possibly in unintended ways. Examples of different types of open worlds problems include the following:

- Usage outside the anticipated physical conditions or environments or performance limits;
- Deployment in systems where unexpected interoperability issues with other applications or systems could arise;
- Users beyond those for which it was designed to be used by and assumptions about them and their understanding and/or motivations for use (e.g., cookies);
- Usage in a wider set of use-cases beyond those for which it was designed;
- Usage in open and/or unregulated environments (as opposed to controlled or regulated environments); and
- Adversarial exploitation.

An example of the first type of problem is the unanticipated conditions encountered by automated systems in automobiles. Reviewing a 2018 incident involving Tesla's Autopilot technology, the National Transportation Safety Board (NTSB) found that the driver "over-relied" on the Autopilot system—the system, which was described as a partially automated system, was used as though it was a fully automated system. The NTSB recommended that automobile manufacturers "limit the use of automated systems to the conditions for which they were designed and ... better monitor drivers to make sure they remain focused on the road and have their hands on the wheel."[38] Similar concerns had been raised in an earlier NTSB report on a 2016 crash between another Tesla and a tractor-semitrailer truck; the report also found that the car's automated control system "was not designed to, and did not, identify the truck crossing the car's path."[39]

An example of the second type stemmed from the use of Bluetooth, a short-range wireless data communications standard that is widely used, including in medical devices. A vulnerability discovered by computer security researchers and known as "SweynTooth" was traced to faulty implementations of the Bluetooth Low Energy protocol, whereby a

---

[38] N. Chokshi, 2020, "Tesla Autopilot Had Role in '18 Crash," *The New York Times*, February 25, 2020, p. B4, https://www.nytimes.com/2020/02/25/business/tesla-autopilot-ntsb.html.

[39] National Transportation Safety Board, 2017, *Collision Between a Car Operating with Automated Vehicle Control Systems and a Tractor-Semitrailer Truck Near Williston*, Florida, May 7, 2016, Accident Report NTSB/HAR-17/02-PB2017-102600, adopted September 12, 2017, https://www.ntsb.gov/investigations/AccidentReports/Reports/HAR1702.pdf.

malformed data packet could result in security breaches and other adverse impacts on the receiving device.[40] When such devices were deployed in the real world, the results were deadlocks, crashes, unpredictable behavior, and the potential for security breaches in the devices. Researchers attributed many of these flaws to inadequate specification of the edge cases, such as handling of partial packets, and inadequate testing in the certification process of the Bluetooth stack. As Kevin Fu observed in his remarks to the committee, the Bluetooth protocols in question were used in a wide array of medical devices.[41] A large number of hospitals and medical offices had to be notified about the need to apply patches to protect patient safety and effectiveness.

Such "open worlds problems" are of relevance to computing researchers whether they are developing systems for real-world deployment or developing methods that others may use for such purposes.

3.2.2 Confronting Cognitive Complexity of Oversight

An approach frequently suggested for handling situations in which computing technologies may err is to recommend human oversight as a remedy, including but not limited to suggesting such oversight to compensate for the biases and limitations of algorithms in criminal justice, work and labor, and health care. This recommendation often takes the form of requiring a human "in the loop" (i.e., engaged in decision-making) or "on the loop" (i.e., monitoring the decision-making). Unfortunately, the burden such oversight places on the people providing oversight is enormous, sometimes one that is impossible to meet. Automated planes provide an interesting and provocative example of how this is so. Captain Chesley Sullenberger correctly noted in a recent interview with *Wired*: "it requires much more training and experience, not less, to fly highly automated planes." The article goes on to observe that "[p]ilots must have a mental model of both the aircraft and its primary systems, as well as how the flight automation works."[42] In a presentation to the committee on computing and civil justice, Ben Green noted that "the vast majority of empirical evidence suggests that people are unable to play the types of roles that ... human oversight and quality control policies imagine." He further argued that "rather than protect against the potential harms of algorithmic decision-making in government, human oversight policies provide a false sense of security in adopting algorithms and enable vendors and agencies to shirk accountability for algorithmic harms."[43]

---

[40] M.E. Garbelini, C. Wang, S. Chattopadhyay, S. Sumei, and E. Kurniawan, 2020, "SweynTooth: Unleashing Mayhem Over Bluetooth Low Energy," Pp. 911-925 in *2020 USENIX Annual Technical Conference (USENIX ATC 20)*, https://www.usenix.org/conference/atc20/presentation/garbelini.
[41] K. Fu, presentation to the committee on May 11, 2021, U.S. Food and Drug Administration.
[42] S. Malmouist, and R. Rapoport, 2021. "The Plane Paradox: More Automation Should Mean More Training," *Wired*, https://www.wired.com/story/opinion-the-plane-paradox-more-automation-should-mean-more-training.
[43] B. Green, "The Flaws of Policies Requiring Human Oversight of Government Algorithms," *Computer Law & Security Review* 45, https://dx.doi.org/10.2139/ssrn.3921216.

Many tasks can be done better by human–computer teams rather than by a person alone or a system alone.[44] The design of systems for such situations must, however, consider human capabilities from the start, and assign to both the person and the system only tasks they are known to be able to handle correctly. Most often doing so will require interdisciplinary work.

3.2.3 Managing Pro-Automation Bias and Automation Aversion

Automation bias and aversion are predictable features of how humans use computing systems, and the conditions in which they occur have been heavily studied. Automation bias refers to people's tendency to defer to (automated) computing systems, leading to their disregarding potentially countervailing possibilities or evidence or failing to pursue them.[45] It has been identified in many different sectors, including aviation,[46] medical care,[47] as well as in the use of computing systems to support the administration of such government functions as welfare, health care, and housing[48] and in the criminal justice system.[49] Renée Hutchins spoke to this point in the context of predictive assessments in the criminal justice system: "We love easy fixes to really complex problems."[50]

Significant ethical and societal impact concerns have arisen with the increased reliance on such technology. For example, tragedy has resulted when too much trust has been given to a semi-autonomous vehicle dubbed an autopilot, and individuals have suffered harms from misplaced reliance on predictions from artificial intelligence (AI) systems developed on biased data. In the other direction, under trust by a user could also result in deadly impacts, if people disable sensor systems that provide warnings because of too many false alarms. As data-intensive applications become more fully

---

[44] B. Grosz, 1994, "Collaborative Systems," *AAAI-94 Presidential Address*, https://aaai.org/Library/President/Grosz.pdf; E. Kamar, S. Hacker, and E. Horvitz, 2012, "Combining Human and Machine Intelligence in Large-Scale Crowdsourcing," In *Proceedings of the 11th International Conference on Autonomous Agents and Multiagent Systems (AAMAS 2012)*; B. Wilder, E. Horvitz, and E. Kamar, 2020, "Learning to Complement Humans," In *Proceedings of the Twenty-Ninth International Joint Conference on Artificial Intelligence (IJCAI-20)*.

[45] R. Parasuraman and V. Riley, 1997, "Human and Automation: Use, Misuse, Disuse Abuse," *Human Factors* 39(2):230-253, https://journals.sagepub.com/doi/10.1518/001872097778543886.

[46] M.L. Cummings, 2004, "Human Supervisory Control of Swarming Networks," Pp. 1-9 in *2nd Annual Swarming: Autonomous Intelligent Networked Systems Conference*, http://hal.pratt.duke.edu/sites/hal.pratt.duke.edu/files/u13/Human%20Supervisory%20Control%20of%20Swarming%20Networks.pdf.

[47] D. Lyell and E. Coiera, 2017, "Automation Bias and Verification Complexity: A Systematic Review," *Journal of the American Medical Informatics Association* 24(2):423-431, https://doi.org/10.1093/jamia/ocw105; R.M. Wachter, 2015, "The Digital Doctor: Hope, Hype, and Harm at the Dawn of Medicine's Computer Age," University of California, San Francisco, https://www.hqinstitute.org/sites/main/files/file-attachments/the_digital_doctor_wachter.pdf.

[48] D.K. Citron, 2008, "Technological Due Process," *Washington Law Review* 85(6), https://openscholarship.wustl.edu/cgi/viewcontent.cgi?article=1166&context=law_lawreview.

[49] K. Freeman, 2016, "Algorithmic Injustice: How the Wisconsin Supreme Court Failed to Protect Due Process Rights in State v. Loomis," *North Carolina Journal of Law and Technology* 18(5), https://scholarship.law.unc.edu/cgi/viewcontent.cgi?article=1332&context=ncjolt.

[50] R. Hutchins, presentation to the committee on March 4, 2021, University of the District of Columbia School of Law.

integrated into people's day-to-day activities, most users may not initially recognize the risks and profound consequences associated with handing over decisions to an intelligent agent.

Automation bias is caused by many different factors. Some of these involve predictable fault on the part of the humans using the system—they choose the path of least cognitive resistance, deferring to the automated system because it is easier to do so than to verify its recommendations; or their role requires oversight of the automated system, and their attention wanders so that they are not properly overseeing it[51] (this is more accurately described as "automation complacency"[52]). Some have to do with the particular affordances of the system, and do not imply human fault. For example, decision support systems that represent their outputs as being highly precise are more likely to be assumed authoritative than if they explicitly represent model uncertainty.[53] And if a decision support system has made a particular recommendation in a high stakes context, such as a decision over pretrial detention or parole, then the human decision-maker knows that, should they overrule the automated system and should its judgment ultimately be vindicated, the human decision-maker is likely to be held accountable. Whereas if they defer to the automated system then the human decision-maker is better able to at least share if not pass on all responsibility. Furthermore, for highly complex decisions, it is understandable that human decision-makers should defer to automated systems, on the grounds that the computer is more likely to have assimilated all of the relevant information than they are.

People also may exhibit algorithm or automation *aversion*: an unwillingness to use reliable algorithms, particularly after seeing the system make a seeming mistake.[54] Algorithm aversion is more likely when people are unable to exhibit any control over the algorithm functioning, or for decisions where they perceive themselves to be (relative) experts or if they are unable to understand the reasons for a system's decisions or actions. Algorithm aversion can also lead to ethically and socially problematic outcomes when useful information or guidance is rejected. At the same time, algorithm aversion

---

[51] L. Bainbridge, 1983, "Ironies of Automation," Pp. 129-135 in *Analysis, Design, and Evaluation of Man-Machine Systems* (G. Johannsen and J.E. Rignsdorp, eds.), Proceedings of the IFAC/IFIP/IFORS/IEA Conference Baden-Baden, Federal Republic of Germany, 27-29 September 1982, https://doi.org/10.1016/B978-0-08-029348-6.50026-9.

[52] R. Parasuraman and D.H. Manzey, 2010, "Complacency and Bias in Human Use of Automation: An Attentional Integration," *Human Factors* 52(3):381-410, https://doi.org/10.1177/0018720810376055.

[53] U. Bhatt, J. Antorán, Y. Zhang, Q. Vera Liao, P. Sattigeri, R. Fogliato, G. Gauthier Melançon, et al., 2020, "Uncertainty as a Form of Transparency: Measuring, Communicating, and Using Uncertainty," Last revised 2021, https://arxiv.org/abs/2011.07586.

[54] J.W. Burton, M.-K. Stein, and T. Blegind Jensen, 2020, "A Systematic Review of Algorithm Aversion in Augmented Decision Making," *Journal of Behavioral Decision Making* 33(2):220-239, https://doi.org/10.1002/bdm.2155; B.J. Dietvorst, J.P. Simmons, and C. Massey, 2015, "Algorithm Aversion: People Erroneously Avoid Algorithms After Seeing Them Err," *Journal of Experimental Psychology: General* 144(1):114, https://repository.upenn.edu/cgi/viewcontent.cgi?article=1392&context=fnce_papers; A. Prahl and L. Van Swol, 2017, "Understanding Algorithm Aversion: When Is Advice from Automation Discounted?" *Journal of Forecasting* 36(6):691-702, https://doi.org/10.1002/for.2464.

typically leads to maintenance of the (potentially problematic) status quo, rather than creation of novel challenges as in the case of automation bias.

3.2.4 Understanding Behavior of Opaque Systems

The inadvertent misuse of any computing system is one consequence of people not understanding the limitations of a system, the reasons it chooses certain actions, or the rationale behind its recommendations. Presentations to the committee in the domains of health care, work and labor, and justice revealed many situations in which such problems arose. In some cases, more robust training of users, with training materials being transparent about the intended use of the systems and the advanced methods they embed and the limitations of those capabilities, will suffice. There are also some methods for building inherently interpretable models. However, most data-intensive AI applications are essentially opaque, "black-box" systems, and new systems capabilities are needed for users to be able to understand the decisions made by the algorithms and their potential impacts on individuals and society.

AI researchers are well aware of these challenges and have begun to explore potential solutions. In particular, research on explainable and interpretable systems is attempting to make it possible to understand the decisions made by such systems well enough to determine their trustworthiness and their limitations given that they do not have access to a predictive model's internal processes.[55] The as yet largely unmet goals of assurance related to explainability are to provide people with the information needed for them to understand the reasoning behind a system's decisions. Interpretable machine learning systems can explain their outputs. For instance, one approach taken toward developing such systems is to consider the degree to which a person can consistently predict the outcomes of such a computing system; the higher the interpretability, the easier it is for a user to comprehend why certain decisions or predictions have been made.[56] Computing research has only begun to address the need for transparency of these systems.

3.3 SOCIETAL CONTEXTS AND DESIGN AND DEPLOYMENT CHOICES

Many of the adverse ethical and societal impacts of computing technology described in the preceding sections result from choices made during design or deployment of a

---

[55] B. Kim and F. Doshi-Velez, 2021, "Machine Learning Techniques for Accountability," *AI Magazine* 42(1):47-52, https://ojs.aaai.org/index.php/aimagazine/article/view/7481; F. Doshi-Velez and B. Kim, 2017, "Towards a Rigorous Science of Interpretable Machine Learning," https://arxiv.org/abs/1702.08608v2.

[56] These terms are used inconsistently in the literature. This explanation differentiates them to emphasize that interpretability is but one approach to explanation.

technology. Some such choices are inherited from the research on which later stages of technology design are based. The first subsection discusses several such choices at the design stage, indicating responsibilities researchers have toward avoiding such consequences and, again, the importance of multidisciplinary collaborations. It also describes challenges of multidisciplinary work and the responsibilities of research sponsors and research institutions toward enabling these kinds of efforts; Recommendations 2, 3, and 4 include specific steps these organizations should take. The second subsection describes ways that researchers and the research community can help those deploying or adopting technologies based on their research make wiser decisions. Practical steps toward these ends are provided by Recommendation 7.

3.3.1 Ideation and Design Stage

Failure to consider a full range of consequences early in the process of developing computing research increases the risk of adverse ethical or societal impacts because researchers have less time to find them and processes long under way (with a variety of investments in the research protocol already undertaken) are more difficult to reform or stop. In addition, even where researchers do manage to consider the right elements late in the research process and are willing to make necessary changes, the costs—intellectual as well as financial—will tend to be higher where the existing research program needs to be drastically changed or even scrapped. Scholarship in the field of design (see Section 3.4.4) has developed theory and methods that enable principled considerations of potential consequences and envisioning alternatives in the design space. This work provides an important foundation for addressing the challenges described in this subsection.

Specifying Intended Functions and Uses of Research and Systems

Computing research is often misdescribed, or even worse, offered with no clear explanation of the appropriate use, function, or domain. Insufficient description is often unintentional, as the researcher herself may not know exactly what application problems the research could assist in addressing. It may even result from good intentions, as the researcher may be trying to avoid biasing others about how the research might be used. Regardless, insufficient descriptions have the potential to lead to adverse behaviors of computing systems that incorporate research outcomes. For example, large language models built to support chatbots, voice assistants, and other language-centric systems are often described as "learning a language" rather than "learning large-scale statistics of word co-occurrence." These misdescriptions may lead to inappropriate research and deployment uses of these language models. For instance, dynamic employee scheduling software is frequently described as "empowering employees" but all too frequently is researched, designed, and developed to empower employers.[57] A related problem arises with rule-based decision systems that are described as aiming to "translate the law into code," when they actually "translate particular legal texts into rules with similar domain."[58] The difference between these goals can be quite important to a potential defendant.

The failure to appropriately specify the intended functions could lead future researchers or deployers to misunderstand the intended functions and uses of the original research and contexts and scopes for which it was developed—and use it in inappropriate ways. Future computing research can be led down dead-end paths and future computing technology development can result in systems that fail in harmful ways. For instance, the description of facial recognition systems using the generic term "computer vision algorithm" implies relative domain- and data-independence. However, the performance of current facial recognition and other perception systems is almost always highly dependent on particular training data sets. The resulting models exhibit biases in that data (e.g., reduced performance on darker-skinned faces).[59]

Researchers should not assume that others (including themselves in the future) will be able to determine or reconstruct the problems that the original research was intended to address. They need to ensure that they have, to the best of their ability, provided enough information that others can appropriately use the results of their research.

Designing Training and Benchmark Data

As Meredith Broussard said,[60] all data are socially constructed. For data sets to be of scientific value and provide a foundation for subsequent research or deployed systems, they need to be intentionally designed and their sample population understood. A number of well-publicized adverse outcomes resulting from data-intensive systems illustrate the problem of bias and coverage inadequacies in training and benchmark data. There are two major sources of coverage inadequacies. One is that there are hard scientific problems to solve in some cases; for example, speech recognition researchers do not know how to handle the challenges of, for example, accent variation. The second is convenience sampling. In 2018, *The Washington Post* worked with two research groups to assess the performance of AI voice assistants against the range of accents present in

---

[57] M.K. Lee, presentation to the committee on March 11, 2021, University of Texas.
[58] B. Barton, presentation to the committee on May 11, 2021, University of Tennessee.
[59] J. Buolamwini and T. Gebru, 2018, "Gender Shades: Intersectional Accuracy Disparities in Commercial Gender Classification," Pp. 77-91 in *Proceedings of Machine Learning Research* 81:1-15, Conference on Fairness, Accountability and Transparency, https://proceedings.mlr.press/v81/buolamwini18a/buolamwini18a.pdf.
[60] M. Broussard, 2018, *Artificial Unintelligence: How Computers Misunderstand the World*, MIT Press, Cambridge, MA.

the U.S. population.[61] The researchers found that for some people with nonnative accents, the inaccuracy rate was 30 percent higher. In another case, biases were discovered in Gmail's Smart Compose feature, which offers suggested text for finishing sentences and for replying to emails. In 2018, a research scientist at Google found that in response to typing, "I am meeting an investor next week," Gmail suggested adding the question, "Do you want to meet him?"—making the assumption that an investor is male.[62] Google's response,[63] removing such suggestions, resulted from an inability to fix the structural gender bias reliably. This example also illustrates the importance of extensive testing before release.

Biased outcomes often result from sampling inadequacies, in particular when predictive functions are developed using data that happens to be available—data gathered from what is called "convenience sampling"—rather than data from carefully designed empirical work or curated with attention to the categories and types of data absent from the data set. Such bias and coverage inadequacies in training and benchmark data typically result from using data collected from online sources or preexisting collections of research study data from conveniently available study participants. For instance, speakers of African American English are today less likely to be present in the research settings where much speech data is collected.[64]

In most cases, these data sets are not the products of purposeful sampling designed to identify the representativeness of the population that generated the data nor do these data sets typically come with metadata to contextualize the social variables that might matter to understanding the data and being able to appropriately apply it to a specific scientific question. For data sets to be of scientific value and provide a foundation for subsequent research or deployed systems, they need to be intentionally designed and their sample population understood. Of course, even carefully designed empirical work can yield data sets that have biases rooted in the biases of the people who participated in the empirical study, particularly if the data set or empirical study in some way assimilates people's biased modes of thought and practices.

An additional coverage problem arises in cases where the data needed are not available in any existing data set. In some cases, that situation results from a population not participating in the activity for which data are being collected or analyzed. This situation arises for minorities in health care. Mays and Cochran noted in their remarks to

---

[61] D. Harwell, 2021, "The Accent Gap," *The Washington Post*, https://www.washingtonpost.com/graphics/2018/business/alexa-does-not-understand-your-accent.

[62] J. Vincent, 2021, "Google Removes Gendered Pronouns from Gmail's Smart Compose to Avoid AI Bias," *The Verge*, https://www.theverge.com/2018/11/27/18114127/google-gmail-smart-compose-ai-gender-bias-prounouns-removed.

[63] A. Caliskan-Islam, J. Bryson, and A. Narayanan, 2016, "Semantics Derived Automatically from Language Corpora Necessarily Contain Human Biases," *Science* 356(6334), https://doi.org/10.1126/science.aal4230.

[64] A. Koenecke, A. Nam, E. Lake, J. Nudell, M. Quartey, Z. Mengesha, C. Toups, J.R. Rickford, D. Jurafsky, and S. Goel, 2020, "Racial Disparities in Automated Speech Recognition," *Proceedings of the National Academy of Sciences* 117(14):7684-7689, https://doi.org/10.1073/pnas.1915768117.

the committee that this problem occurs especially for data that involves intersectional characteristics—that is, when multiple axes of disadvantage or underrepresentation are present. In other cases, there may be ethical questions related to collecting the data.[65] For example, Amy Fairchild discussed in her remarks to the committee the challenges of conducting HIV surveillance in a country that criminalizes homosexuality. She observed that such decisions have implications for a group's power to advocate or seek resources and that decisions about how such information is used must be made in consultation with the community members who would be most affected.[66]

The data may also interact with other system features (e.g., algorithm or objective function choice) to yield biases. For example, a shopping algorithm may be designed to vary its output based on information about shopping behaviors. This information might inadvertently correlate directly to gender or race (even if that specific data label is not fed into the training algorithm), so that the resulting predictive system becomes a biased decision-making vehicle. Furthermore, the complexity of these algorithms can make it almost impossible to validate the systems or understand their results (see the subsection "Validation" below), making it likely that the harms they engender will outweigh their benefits.[67]

In her remarks to the committee, Sarah Brayne noted that "[the] premise behind predictive policing algorithms and the training data [is] that you can learn about the future from the past. And so, any inequality in the historical data is going to be reflected and projected into the future."[68] Ben Green observed in his remarks to the committee that "given existing racial and other disparities in outcomes such as creditworthiness, crime risk, educational attainment, and so on, even perfectly accurate predictions would reproduce social hierarchies. Striving primarily for more accurate predictions of outcomes may enable public policy to naturalize and reproduce inequalities."[69]

Illustrating the real consequences, the harm such systems can cause was described by Renée Hutchins, "recent data comparing Black and White stop and arrest rates suggest that you are twice as likely to be arrested if you're Black and five times more likely to be stopped without cause. And while stops and arrests may ultimately be shown to be unconstitutional within the criminal justice system, in the interim, they are fed into data modeling that is used for future predictions about criminality."[70] The proliferation of sensors throughout society has led to increased numbers of people who have had no police contact being included in law enforcement databases.

---

[65] V. Mays and S. Cochran, presentation to the committee on May 6, 2021, University of California, Los Angeles.

[66] A. Fairchild, presentation to the committee on March 16, 2021, The Ohio State University.

[67] A. Howard, 2019, "Demystifying the Intelligence of AI," *MIT Sloan Management Review*, https://sloanreview.mit.edu/article/demystifying-the-intelligence-of-ai.

[68] S. Brayne, presentation to the committee on March 4, 2021, The University of Texas at Austin.

[69] B. Green, 2022, "Escaping the Impossibility of Fairness: From Formal to Substantive Algorithmic Fairness," Last revised April 26, http://dx.doi.org/10.2139/ssrn.3883649.

[70] R. Hutchins, presentation to the committee on March 4, 2021, University of the District of Columbia School of Law.

Computing researchers are aware of these issues and are making efforts to address them, through developing techniques for correcting biased data, through developing different algorithms for learning, and through collaborative research with subject domain experts. For example, computing researchers working with those with legal expertise may be able to help mitigate bias. Jens Ludwig noted that arrests for lower-level offenses are more subject to discretion, and hence bias, than arrests for more serious offenses, and that the court system and convictions are less prone to bias than arrests.[71] "So, we built a tool that focuses on using convictions for relatively more serious offenses, ignoring less serious offenses, and the result is you can see a tool that gives almost identical release recommendations [for Blacks and Whites]."[72] It is crucial that efforts addressing bias engage social science and ethics expertise as they involve applying a variety of nuanced social scientific concepts (as Section 2.2 describes).

Defining Objective Functions

Data-intensive machine learning methods maximize (or minimize) some objective function during training. There are many types of objective functions, including cost and loss functions that evaluate how well a specific search algorithm models a given set of data. The selection of objective functions significantly influences what is learned. They reflect those values the designer considers important to the decisions or predictions the system will make.

Several presentations to the committee pointed to cases in which the choice of objective function resulted in outcomes that favored one group over another. For instance, the optimizations of algorithmic management systems may omit certain factors important to workers: when developing shift work schedules, management may prioritize workplace efficiency and economic value while attention to worker well-being might prioritize stability and consistency of a schedule.[73] Participatory design approaches can enhance worker well-being.[74] When learning ways to identify a good worker, the objective function may incorporate focus on things that are easy to count, like the number of

---

[71] J. Ludwig, presentation to the committee on March 4, 2021, University of Chicago.

[72] For example, previous studies such as Mitchell and Caudy (2013) use survey data to estimate the probability of arrest for low-level offenses, such as drug charges, conditional on self-reported involvement in such offenses and find large disparities by race in arrest likelihood. In contrast, studies such as Beck and Blumstein (2018) find that racial disparities in sentencing outcomes are much smaller in proportional terms conditional on current charge and prior record, especially for the most serious offenses. A.J. Beck, and A. Blumstein, 2018, "Racial Disproportionality in US State Prisons: Accounting for the Effects of Racial and Ethnic Differences in Criminal Involvement, Arrests, Sentencing, and Time Served," *Journal of Quantitative Criminology* 34(3):853-883. O. Mitchell and M.S. Caudy, 2015, "Examining Racial Disparities in Drug Arrests," *Justice Quarterly* 32(2):288-313, https://doi.org/10.1080/07418825.2012.761721.

[73] M.K. Lee, presentation to the committee on March 11, 2021, The University of Texas.

[74] M.K. Lee, I. Nigam, A. Zhang, J. Afriyie, Z. Qin, and S. Gao, 2021, "Participatory Algorithmic Management: Elicitation Methods for Worker Well-Being Models," Pp. 715-726 in *Proceedings of the 2021 AAAI/ACM Conference on AI, Ethics, and Society,* Association for Computing Machinery, New York, https://doi.org/10.1145/3461702.3462628.

email messages responded to or number of lines of code without knowing first whether such measures have a positive correlation with productivity rates.[75] Ece Kamar noted that AI systems (research and development [R&D]) typically are optimized for fully automated work, assessing accuracy as if systems are going to be working alone; teamwork is not part of such optimizations.[76] Changing the objective function to a team centric view enables prioritizing learning about things that humans are not very good at. And doing that yields performances that are better than having either the computing system or the human doing tasks alone.[77]

Extreme risk scenarios can arise from an insufficiently thought-out objective function. For example, systems designed to maximize the defense and protection of military assets that fail to consider how their adaptive behaviors could affect the risk of war and systems designed to optimize electric grid efficiency can exacerbate cybersecurity risk.

The choice of the objective function, as well as such other key components of a learning system as the algorithm to select and the random seed function are key elements of the design process. For example, ensemble methods, learning systems that combine different kinds of functions each with their own different biases, have long been credited as often performing better than completely homogeneous methods.

Engaging Relevant Stakeholders

The outcomes of computing research may be directly integrated into deployed systems or inform their design. As a result, the inclusion of the interests, values, and needs of a variety of stakeholders at the earliest stage of computing research becomes important not only for system success but also for alerting fellow researchers and society about potential limitations and concerns. The more obvious stakeholders of computing research are the "end-users" who use artifacts and other research products, but as noted earlier, there are many others.

In the case of algorithmic systems supporting pretrial release decisions, defendants and not just judges and prosecutors are stakeholders. Defendants are not "users" of those systems in the traditional sense but are stakeholders because their lives will be profoundly impacted by how the system behaves—that is, by the algorithmic design. Even though the research itself (e.g., algorithm design or development of a new HCI method) may not directly involve all stakeholders, lack of attention to the values and needs of the wider communities affected by the systems the computing research enables may nevertheless have adverse outcomes for them. Most often, it is these neglected or overlooked

---

[75] K. Levy, presentation to the committee on March 11, 2021, Cornell University.
[76] E. Kamar, presentation to the committee on March 11, 2021, Microsoft Research.
[77] B. Wilder, E. Horvitz, and E. Kamar, 2020, "Learning to Complement Humans," in *Proceedings of the Twenty-Ninth International Joint Conference on Artificial Intelligence (IJCAI-20)*.

stakeholders who incur the greatest risks; with some forethought and attention, responsible computing research and the technologies that follow from it can reduce the risks that computing technology will introduce more harm than social good.

The need to engage the full spectrum of stakeholders may be most pronounced when technological solutions are sought for social problems, because "technological solutionism" often involves prioritizing the needs of or directing resources to private actors without adequate community involvement or democratic oversight.[78,79] The values and interests of people and groups who are not well represented in computing research are at particular risk of being systematically ignored. In the absence of rigorous methodologies and frameworks for identifying the complicated social dynamics (outlined earlier in the report) that shape the problems that computing research strives to address, computing research is much less equipped to produce theories, products, or artifacts, not to mention deployed systems into which that research feeds, that adequately solves for those most in need of what computing has to offer.

Panel presentations by health care experts illuminated the importance of engaging stakeholders by highlighting the striking contrast between the non-involvement of clinical staff in electronic medical record design and their very successful incorporation into the design of an early warning system for sepsis. The key difference in the sepsis research outcomes came from engaging nurses to understand their expert knowledge of handling of sepsis and the workflows that prove critical to them executing that expertise. An interdisciplinary team of researchers developed Sepsiswatch so that it could track the stakeholders of the current approaches and systems built to monitor patient infections and understand who benefits the most from the current workflows. From there, computing researchers, informed by social scientists on the team, came to their designs with a clearer sense of how to distribute the risks and benefits of a system for monitoring patient outcomes as equitably as possible. A more effective, responsible mechanism for sepsis management in a clinical setting resulted from including the individuals who regularly contributed to patient monitoring, as well as a broader group of stakeholders including the patients, nursing staff, and hospital administration. Panelists also spoke about the challenges of engaging stakeholders in a way that crystallizes their needs, conveying constraints and possibilities to systems designers.[80]

Another example from work and labor of the importance of computing research considering a fuller range of stakeholders in its designs comes from a different area of health care. In the 1990s Kaiser Permanente began developing a robotic system to

---

[78] S. Brayne, 2021, *Predict and Surveil: Data, Discretion, and the Future of Policing*, published to Oxford Scholarship Online, Oxford University Press, United Kingdom, https://oxford.universitypressscholarship.com/view/10.1093/oso/9780190684099.001.0001/oso-9780190684099.

[79] B. Green, 2019, *The Smart Enough City: Putting Technology in Its Place to Reclaim Our Urban Future*, MIT Press, Cambridge, MA.

[80] L. Sweeney, presentation to the committee on March 16, 2021, Harvard University.

assist environmental service workers in cleaning a hospital. Engineers engaged these workers alongside medical providers in the design. The environmental service workers' knowledge of practical ways to combat infection and bacteria in rooms made the system design better than if the engineers had only talked with hospital clinical staff or administrators. Presentations to the committee on labor and work also provided examples that illustrated the problems of not engaging stakeholders: fast food workers left out of a design for food safety could not use a system that did not understand their everyday workflows;[81] groundskeepers had to contend with the noise and disconcerting worker surveillance of a drones system designed to help landscaping efforts that did not assume people's work might be made less productive by the drone's presence,[82] and dynamic scheduling of work schedules negatively impacted the well-being of shift workers because these systems assumed work assignments were the most important factor while workers needed to prioritize other life demands such as commutes and childcare.[83]

Technical computer science and computing research training does not currently provide computing researchers with the knowledge and skills needed to move beyond the instinct to develop new technologies that they imagine would be terrific for themselves or people with whom they regularly interact. Nor are there incentives for computing researchers to draw on social scientific expertise to identify and engage stakeholders to better map out the social dynamics that could inform a system's design. For instance, surveillance cameras might make janitors on the night shift feel safer or make them feel as if they are being surveilled.[84] Different choices of algorithms may lead to different ways of balancing the trade-off between these two likely effects. To know how to think about such eventual outcomes, computing researchers recognize the importance of thinking about the ways a new technology (incorporating or based on their research) might be used, by whom, and in what contexts and with what potential impacts. It is unrealistic to expect all computer scientists to develop such expertise, but they should appreciate its importance and learn how to work with those with such expertise. The successful development of all the systems discussed in presentation to the committee had one thing in common: they involved computing researchers incorporating the insights and subject-matter expertise of a range of stakeholders who were not obvious end-users of their systems. In most cases, these success stories involved computing researchers working with social scientists trained to see the stakeholders in the mix. Stakeholders are sometimes obvious. More often, they are groups or individuals harder to see if one is tightly focused on who might buy or use a piece of technology, or worse, if one assumes that it does not matter who might use a system.

---

[81] M.K. Henry, presentation to the committee on April 29, 2021, Service Employees International Union.
[82] Ibid.
[83] M.K. Lee, presentation to the committee on March 11, 2021, The University of Texas.
[84] M.K. Henry, presentation to the committee on April 29, 2021, Service Employees International Union.

Integrating Computing and Domain Expertise

Computing systems are increasingly essential infrastructure for other disciplines as well as having impact across much of daily life. For them to work well requires expertise in the domain of application (see also Section 3.1.5, "Aligning with Existing Norms, Structures, and Practices"). It also requires expertise in the social sciences that is important both for bridging and synthesizing across a range of subject-matter expertise. For instance, they can provide expertise in ways to understand the distribution of risks and benefits as well as to resolve tensions among stakeholders. Recent work on epidemiological modeling, projects in linguistics and large language models, and contact tracing have all shown the importance of engaging with domain experts. Additional challenges arise when data-driven systems are used for advocacy by multiple parties who may be competing or engaged in an adversarial decision process such as those used in the legal system.

As computing researchers as well as researchers and scholars in other disciplines have limited expertise, designing systems that work well requires a partnership between computing and domain experts. Absent such a partnership, systems typically fail. For example, Robert Wachter pointed out that the "battle" for becoming the dominant EHR company was not won by any of the leading companies that originally tried (including IBM, General Electric, Google, and Microsoft), because they lacked sufficient health care domain knowledge and focus.[85] Instead, the two leading EHR vendors were companies built solely for the purpose of creating and selling EHRs: Epic and Cerner. That domain knowledge proved more crucial than competencies in data analytics, artificial intelligence, data visualization, and consumer facing cloud tools. The experience of M.D. Anderson with IBM Watson on cancer therapy recommendations efforts is another notable example.[86]

There are many notorious examples of poor outcomes when computer science researchers work without regard to the bodies of knowledge in other disciplines,[87] demonstrating the importance of interdisciplinary partnerships to responsible computing research. One example of a true and successful interdisciplinary research partnership is the decryption of the Copiale Cipher through a collaboration between a computer scientist

---

[85] R. Wachter, presentation to the committee on March 16, 2021, University of California, San Francisco.
[86] E. Strickland, 2019, "How IBM Watson Overpromised and Underdelivered on AI Health Care," *IEEE Spectrum*, https://spectrum.ieee.org/how-ibm-watson-overpromised-and-underdelivered-on-ai-health-care.
[87] J. Dron, ed., "Bad Research and Practice in Technology Enhanced Learning," *Education Sciences* (ISSN 2227-7102), https://www.mdpi.com/journal/education/special_issues/technology_education; T. Evgeniou, D.R. Hardoon, and A. Ovchinnikov, 2020, "What Happens When AI Is Used to Set Grades?" https://hbr.org/2020/08/what-happens-when-ai-is-used-to-set-grades; D. Coldewey, 2020, "Google Medical Researchers Humbled When AI Screening Tool Falls Short in Real-Life Testing," *TechCrunch*, https://techcrunch.com/2020/04/27/google-medical-researchers-humbled-when-ai-screening-tool-falls-short-in-real-life-testing; W.D. Heaven, 2021, "Hundreds of AI Tools Have Been Built to Catch COVID. None of Them Helped," *MIT Technology Review*, https://www.technologyreview.com/2021/07/30/1030329/machine-learning-ai-failed-covid-hospital-diagnosis-pandemic.

and two linguists.[88] The success of this effort led in turn to a long-term research project involving computer scientists, linguists, and historians.[89]

Interdisciplinary involvement in many areas of computing research requires a collaboration of computing researchers and disciplinary experts as equals. Too often, interdisciplinary research that involves computer scientists devolves into a "consultant" model—either the computer scientists are treated as software developers or the researchers from other disciplines are minimally included. For example, an analysis of projects funded under the National Science Foundation's Information Technology Research program found that nearly one-third of senior personnel on these projects did not publish together.[90]

Interdisciplinary research projects are, however, more difficult to conduct; it is necessary for the collaborators to develop understanding of the terminologies, concepts, and methods of each discipline, and this takes time.[91] As discussed in the subsection "Integrating Ethical and Societal Issues into Training," earlier in this chapter, a broader education than is the current standard in computer science is needed and can start at the undergraduate level. Furthermore, the structure of academia and academic promotion processes continues to inhibit the formation of such partnerships. Interdisciplinary research may be disregarded as part of tenure and promotion; different fields value different types of research productivity (e.g., conference versus journal publications) and non-core computing research may be considered soft. As Madeleine Clare Elish pointed out, "it doesn't count for tenure to ... work in new spaces."[92] Ben Green remarked that "it's very hard at universities to actually create these types of deeply integrated interdisciplinary environments" and pointed out the need for "funders to create mechanisms for actually doing that."[93] Thus, research organizations and scientific and professional societies need to adapt their structures and evaluation processes, so they properly recognize such research.

3.3.2 Deployment

Characteristics of computing systems design and the information provided about system capabilities influence decisions made by those deploying new technologies and can

---

[88] J. Markoff, 2011, "How Revolutionary Tools Cracked a 1700s Code," *The New York Times*, https://www.nytimes.com/2011/10/25/science/25code.html.

[89] Swedish Research Council, "Automatic Decryption of Historical Manuscripts: The DECRYPT Project," https://de-crypt.org.

[90] J.N. Cummings and S. Keisler, 2008, "Who Collaborates Successfully? Prior Experience Reduces Collaboration Barriers in Distributed Interdisciplinary Research," Pp. 437-446 in *Proceedings of the ACM Conference on Computer Supported Cooperative Work*, CSCW, http://dx.doi.org/10.1145/1460563.1460633.

[91] E. Brister, 2016, "Disciplinary Capture and Epistemological Obstacles to Interdisciplinary Research: Lessons from Central African Conservation Disputes," *Studies in History and Philosophy of Biological and Biomedical Sciences* 56:82-91, https://doi.org/10.1016/j.shpsc.2015.11.001.

[92] M.C. Elish, presentation to the committee on March 16, 2021, Google, Inc.

[93] B. Green, presentation to the committee, May 25, 2021, University of Michigan.

affect the societal impact of deployed systems. Although deployment is downstream from computing research, researchers and the computing research community incur responsibilities related to enabling acquirers of new technologies to make wise decisions. For them to meet these responsibilities requires their taking into account various features of deployment. This section describes three sources of potential ethical and societal impact concern: institutional pressure on procurement of technologies to address societal problems, challenges presented by the complex nature and development of computing systems, and challenges of ensuring appropriate use. It also discusses challenges of disparate access to new technologies and the importance of governance mechanisms and regulation. Recommendations 7 and 8 include practical steps researchers can take to help address these concerns. Recommendation 3.4 indicates steps academic institutions can take in educating students to help.

Acknowledging Institutional Pressures

Some of the factors that drive organizations to deploy new computing technologies can lead to problematic outcomes either for those organizations or for individuals or groups affected by the actions and decisions of those organizations. Presentations to the committee relating to the use of computing technology in the public sector revealed three types of challenges for those making acquisition decisions: (1) pressures to improve the efficiency of the organization, (2) pressures to improve the accountability of the organization, and (3) insufficient knowledge in institutions about the technologies they are considering procuring. In each case, there may be opportunities for computing researchers to help such institutions in making better decisions.

Institutions under pressure to enhance efficiency will sometimes turn to computing technologies even if the case that they will in fact yield greater efficiency has not been made or the groundwork needed to realize those benefits has not been laid or sufficient attention given to the fact that technological approaches alone cannot solve societal problems (see Section 3.1.4, "Proper Roles for Technologies in Addressing Societal Problems"). Discussing adoption of predictive analytic policing tools by the Los Angeles Police Department, for example, Sarah Brayne[94] observed that the tools appeared to have been adopted not necessarily because there was empirical evidence that their use would actually improve outcomes of interest but rather because the department, like many other government agencies, was facing institutional pressures to adopt data analytics. These pressures arose from an impression that their use would result in more efficient allocation of law enforcement resources as well as improve objectivity and reduce bias in the department's decision-making.

---

[94] S. Brayne, presentation to the committee on March 4, 2021, The University of Texas at Austin.

Health care delivery provides another example: a desire to improve the efficiency and quality of U.S. health care prompted the federal government to adopt incentives and penalties for hospitals and medical offices to adopt EHRs. As the systems were rolled out, both the medical practitioners who used them and the institutions that deployed them came to understand that merely digitizing health records was far from sufficient to achieve the efficiency and quality goals. Much work would be needed to realize the vision of EHRs and to understand and transform the work, workflows, and relationships associated with delivering medical care needed to take full advantage of those EHRs.

In his remarks to the committee,[95] Ben Green observed that "algorithmic reforms are simultaneously too ambitious and not ambitious enough. On the one hand, algorithmic interventions are remarkably bold: algorithms are expected to solve social problems that couldn't possibly be solved by algorithms. On the other hand, algorithmic interventions are remarkably timid and display a notable lack of social or political imagination: such efforts rarely take aim at broad policies or structural inequalities, instead opting merely to alter the precise mechanisms by which certain existing decisions are made."[96]

Government agencies and other institutions also face pressures to use computing technologies in an attempt to improve accountability. For example, Sarah Brayne described how the Los Angeles Police Department responded to a consent decree by deploying a new data-driven employee risk management system and associated capabilities for data capture and storage that subsequently raised a set of societal issues because of mission creep.[97] (See the subsection "Mission, Function, and Scale Creep" below.)

Last, government agencies and other institutions frequently lack the in-house technical expertise to make informed decisions about the design and implementation of computing technologies. This is exacerbated when developers of technology are unclear about a technology's limitations. In his presentation to the committee, Jens Ludwig noted that Broward County, Florida, might routinely procure millions of dollars of laptop computers and cell phones, able to draw on quality information that is widely available to consumers. Ludwig contrasted this with the county's procurement of the Correctional Offender Management Profiling for Alternative Sanctions (COMPAS) case management and decision support tool used to assess the likelihood of recidivism. Ludwig concluded that it "would be fair to wonder to what degree COMPAS was evaluated prior

---

[95] B. Green, presentation to the committee, May 25, 2021, University of Michigan.
[96] B. Green, "Algorithmic Imaginaries: The Political Limits of Legal and Computational Reasoning," Law and Political Economy Blog, 2021, https://lpeproject.org/blog/algorithmic-imaginaries-the-political-limits-of-legal-and-computational-reasoning.
[97] S. Brayne, presentation to the committee on March 4, 2021, The University of Texas at Austin.

to deployment by Broward County in terms of its accuracy ... and fairness in Broward County."[98]

At the same time, vendors frequently consider the detailed working of their systems to be proprietary. In his remarks to the committee, Dan Ho cited as an example a U.S. Customs and Border Protection procurement of biometric systems for use in border entry. The agency's efforts to identify the cause of failure with an iris scanning system were stymied by the inability to understand proprietary technology. When contractors do not provide additional detail, according to Ho, the agency's ability to oversee the program can be undermined.

The lack of in-house technical knowledge and vendors' claims that their technologies are proprietary lead to what economists refer to as a principal agent problem,[99] in which vendors know a great deal more about a system's performance, limitations, and shortcomings than the acquirer does. This informational asymmetry can have significant effects outside of the acquiring institution. With systems that make consequential decisions such as where to focus policing or whether to grant bail, there can be significant negative consequences for individuals and for groups in the communities the institutions serve.

Ensuring Appropriate System Characteristics

Continuous integration and continuous deployment. The pace of innovation in computer science is rapid, which is both a blessing and a curse. In the best case, low barriers to implementing new ideas enable real problems to be solved quickly and efficiently. However, in the worst case, the urge to immediately release "the next big thing" leads to a reckless disregard for downside risk, and to a "building for building's sake" mentality that deprioritizes the fundamental goal of R&D: to generate new insights and new technologies that serve a higher societal purpose. Compounding the problem is the popular practice of continuous integration and continuous deployment (CI/CD) in which a tech product is expected to have flaws throughout its deployment, and to receive a constant stream of tweaks along the way. Kevin Fu observed that testing before deployment is understood in the health care setting to be life critical: "even if the

---

[98] COMPAS was purchased by Broward County, Florida, in 2008. Prior to that point, evaluations of COMPAS by Northpointe (the software's creator) had apparently been limited to other jurisdictions, specifically parole systems in New York and California. In 2009 the Broward County auditor's office published an evaluation of the Pretrial Service Program in the county (E.A. Lukic, 2009, *Evaluation of the Pretrial Services Program Administered by the Broward Sheriff's Office*, Report No. 09-07, https://www.broward.org/Auditor/Documents/pretrial_final060909.pdf) noting that the COMPAS tool had by then still not been validated in Broward County specifically. The first validation of COMPAS in Broward County specifically that is publicly available seems to be a 2011 analysis by Florida State University (Florida State University College of Criminology and Criminal Justice, 2011, "Managing Broward County's Jail Populations: Validation of the COMPAS Risk Assessment," https://criminology.fsu.edu/sites/g/files/upcbnu3076/files/2021-03/Broward-COMPAS-Validation.pdf).

[99] J. Ludwig, presentation to the committee, March 4, 2021, University of Chicago; J.-J. Laffont and J. Tirole, 1993, *A Theory of Incentives in Procurement and Regulation*, MIT Press, Cambridge, MA.

software patch is available, it might not be deployed overnight" because it takes time to analyze the impact of a patch on behavior of the overall system.[100]

The CI/CD model is seductive because it might be seen as absolving technologists of the burden of forethought; post-deployment problems are seen as inevitable, and as an acceptable cost of progress. However, the costs of addressing problems after the fact is often much higher than the cost of addressing them during the initial design of a system. Indeed, it is considered good industrial practice for each system update to go through multiple review steps for quality, safety, privacy, and reliability.

Sometimes, however, research teams may release experimental systems with less scrutiny. For example, the Microsoft Tay chatbot was released directly from an R&D group to the public, and within 24 hours had been trained to be sexist and racist.[101] Ece Kamar commented that there is an "organizational question for anybody who is deploying these complex computational systems in their workplaces, how to ensure that there are feedback loops."[102] CI/CD reduces the friction of deploying changes to market but risks the devaluation of substantive (albeit slower moving) discussions about the unexpected consequences of a change. Put another way, there is a tension between the natural desire of researchers to test their latest idea "in the wild" and institutional processes that rank values such as safety and privacy more highly.

Validation. Why should one have confidence in a computational system? Why should it be entrusted with sensitive data, or the ability to make decisions with important consequences in the real world? In part, the ability to trust a system is derived from characteristics that lend themselves to objective definitions and measurements. One can measure how often a system crashes, and how many seconds it needs to handle a request; one can verify that the system has installed up-to-date security patches and protects data with encryption and access controls. However, a system that is secure, fast, and highly available may still produce results that humans (or other systems) should not trust. So how should one define the *validity* of a system's results? In other words, how should one evaluate our confidence that the results are appropriate reflections of the ultimate goals of the system?

These computer-systems questions are related to the instrumental ethical values of trustworthiness, verifiability, assurance, and security described in Section 2.1. Answering the systems validity question is increasingly difficult for the complex systems of the modern era. These systems are typically built without a formal specification that produces

---

[100] K. Fu, presentation to the committee on May 11, 2021, U.S. Food and Drug Administration.
[101] M.J. Wolf, K.W. Miller, and F.S. Grodzinsky, 2017, "Why We Should Have Seen That Coming: Comments on Microsoft's Tay 'Experiment,' and Wider Implications," *The ORBIT Journal* 1(2):1-12, https://www.sciencedirect.com/science/article/pii/S2515856220300493.
[102] E. Kamar, presentation to the committee on March 11, 2021, Microsoft Research.

rigorous, comprehensive test cases by which concrete implementations can be evaluated. For example, the high-level goals of an operating system (OS) such as Linux are sufficiently well understood to enable independent groups of developers to work on different parts of the OS in parallel, with the pieces eventually integrating to work together in a (mostly) cohesive whole. However, an OS in practice is sufficiently complicated that emergent problems can and do occur, with the appropriate solution often requiring subjective reasoning. For example, what should happen when a change in the OS's scheduling algorithm, introduced to make certain workloads run faster, has a detrimental effect on other ones?[103] Even though the Linux community has a variety of tools for determining how new kernel updates impact objective performance metrics,[104,105] many updates involve trade-offs between different metrics, requiring human reasoning to decide if the net result is positive.

The challenge of defining validity exists in every subdiscipline of computer science, but the rise of machine learning has provided salient examples. For instance, the oft-lamented problem of biased training data is really a problem of validity. A good example is how distributional biases in associating professions with gender in training data lead to not just biased, but actually incorrect translations.[106]

Another illustration comes from the problem of face recognition. An image database of human faces, used to train a facial recognition algorithm, is invalid if the data set lacks demographic diversity. From a narrow mathematical lens, the invalidity can be defined with respect to the statistical divergence of the data set's images from the richness of faces that exist in real-life. However, this statistical notion of invalidity is fundamentally given meaning by a *values-based decision* that humans must make—the statistical divergence is only important because we as humans should prefer to live in a society where facial recognition technology works equally well for all people, regardless of characteristics like age, gender, or race. Thus, defining "validity" in computer science requires something beyond mere technical skill; it requires moral imagination.

Mitchell Baker of Mozilla drew attention to yet another kind of validation challenge—the difficulty of academic researchers being able to investigate the ethical and societal implications of today's Internet-scale computing systems.[107] To do so would require that the researchers have access to the large data sets these systems rely on, the models built with that data, and large-scale computational platforms.

---

[103] D. Chiluk, 2019, "Unthrottled: How a Valid Fix Becomes a Regression," *Indeed Engineering Blog*, https://engineering.indeedblog.com/blog/2019/12/cpu-throttling-regression-fix.

[104] T. Chen, L.I. Ananiev, and A.V. Tikhonov, 2007, "Keeping Kernel Performance from Regressions," *Proceedings of the Linux Symposium, Volume One,* Ottawa, Ontario, Canada, https://www.kernel.org/doc/ols/2007/ols2007v1-pages-93-102.pdf.

[105] Intel, 2021, "Linux Kernel Performance Project," https://01.org/lkp.

[106] K. Webster and E. Pitler, 2020, "Scalable Cross Lingual Pivots to Model Pronoun Gender for Translation," arXiv, https://doi.org/10.48550/arXiv.2006.08881.

[107] M. Baker, presentation to the committee on June 24, 2021, Mozilla Corporation.

However, large data sets of great interest to researchers, such as those containing user-generated content and user interactions, is highly concentrated among a handful of companies. Some of these firms have developed mechanisms to share data with some academics. Initiatives such as the National AI Research Resource aim to widen access to computational resources and data so that a broader swath of academia will be able to carry out research in this arena. However, as discussed further at the end of Section 3.5.2, there are manifest tensions among the interests of researchers, the proprietary interests of companies, and the privacy interests of users.

Last, validation also requires the courage to admit that our assumptions may be incorrect. For example, the idea that "more data leads to better decision-making" is intuitively appealing. However, algorithmically generated decisions do not automatically lead to better outcomes. As Dan Ho described in his remarks to the committee,

> Beginning in the 1990s, criminologists advanced predictive policing as a method to forecast crime hotspots and drive down crime. Several jurisdictions conducted rigorous evaluations and leading studies showed no benefit in terms of crime reduction.[108] We should not underestimate the value of rigorous inquiry; if only limited parties have access to the data to evaluate systems, accountability is not going to be possible.

Ensuring Appropriate System Use

Mission, function, and scale creep. Computing technologies are typically developed and deployed to address particular needs or challenges, and a deployed computing technology can have many different intended missions, functions, or scales at which it is intended to operate, where some might be unstated or implicit. Because computers are universal machines, part of their power is that computing technologies developed for one purpose might be used for a range of other purposes. All is well and good if the new use is an appropriate one.

In general, however, technologies—including computing technologies—that are developed and deployed for one function might be inappropriate, or even harmful, for other functions. In many cases, significant challenges arise when the mission, function, or scale change over time, particularly if the changes are not explicitly noted (as there is "creep" of various sorts). For example, when used by small groups, over a short time, or

---

[108] P. Hunt, J.M. Saunders, and J.S. Hollywood, 2014, *Evaluation of the Shreveport Predictive Policing Experiment*, RAND Corporation, Santa Monica, CA; J. Saunders, P. Hunt, and J.S. Hollywood, 2016, "Predictions Put into Practice: A Quasi-Experimental Evaluation of Chicago's Predictive Policing Pilot," *Journal of Experimental Criminology* 12:347-371, https://doi.org/10.1007/s11292-016-9272-0. Ho also indicated that some studies did find effects: G.O. Mohler, M.B. Short, S. Malinowski, M. Johnson, G.E. Tita, A.L. Bertozzi, and P.J. Brantingham, 2015, "Randomized Controlled Field Trials of Predictive Policing," *Journal of the American Statistical Association* 110(512):1399-1411, https://doi.org/10.1080/01621459.2015.1077710.

with limited reach, algorithms for collaborative filtering of news items can help people quickly access information that is more likely to be relevant to their current needs. However, there are concerns that when deployed at global scale those same algorithms can in some cases contribute to the creation of echo chambers, information bubbles, and increased polarization. Presentations to the committee in health care, work and labor, and justice discussed several others, including

- Workforce monitoring technologies can be beneficial in terms of identifying potential safety risks, but their mission can easily creep into harmful and invasive surveillance.
- Many EHR systems work well for information recording and transfer, particularly for billing purposes. However, they are now also expected to support the functions of health care delivery, including diagnosis and planning. Numerous studies have shown that EHR systems create numerous challenges and health care failures, in part because they are being used for different functions for which they were not appropriately designed.
- As Sarah Brayne described in her remarks to the committee, the Los Angeles Police Department deployed a new data-driven employee risk management system and associated capabilities for data capture and storage as part of a consent decree with the Department of Justice. Intended to improve accountability, it spurred, according to Brayne, a proliferation of automated decision-making throughout the department and the repurposing of data.[109]

Significant societal impact problems can arise if computing methods are deployed at a larger scale than originally intended. Systems that work at a small scale may not work at a large scale as Twitter learned that the hard way during the 2010 World Cup when the increase in tweets per second led to short periods of unavailability.[110] Such experiences point to the conflicts between moving quickly in a competitive environment and following good engineering practice. For computing researchers, two key issues are how to create experimental frameworks that facilitate safe staged deployment and how to teach researchers to accept a slower pace to their impact in service of greater care in avoiding unintended consequences.

These kinds of "creep" clearly present significant issues for those deploying computing technology, and better practices around implementation can help guard against problematic creep. But the possibility of problematic creep also raises issues for

---

[109] S. Brayne, presentation to the committee on March 4, 2021, The University of Texas at Austin.
[110] J. Reichhold, D. Helder, A. Asemanfar, M. Molina, and M. Harris, 2013, "New Tweets per Second Record, and How!," *Twitter*, https://blog.twitter.com/engineering/en_us/a/2013/new-tweets-per-second-record-and-how.

computing researchers. Appropriate development and deployment require an understanding of the capabilities of the underlying computing technology. Moreover, research and deployment often occur in a distributed or decentralized manner, so that no individual or group is involved in every step from research idea to final implementation. Although computing researchers cannot possibly anticipate every possible use of their work, they have an obligation to be clear about the exact functionality and appropriate uses of the products of their research (see also the subsection "Specifying Intended Functions and Uses of Research and Systems" earlier). Doing so will help those using their research results to make better informed decisions about what missions, functions, or scales are appropriate for a computing technology.

Strategic behavior by individuals and institutions. Negative societal outcomes from computing systems may also come from a failure to adequately anticipate strategic behavior by end-users or by other computing systems. Strategic behavior can take many different forms. Most commonly, a computing system is designed with the goal of optimizing for some property, which is not directly observable by the system. The system therefore optimizes for some measurable feature. Often that feature diverges from the goal by enough that, when users learn what is being optimized for, they are able to manipulate the system to receive preferred outcomes without actually better exemplifying the underlying property for which the system is intended to optimize. Examples of this abound, including search engine optimization (heaps of links), students gaming the computer grading of papers, and consumers manipulating their FICO scores.[111] There are some notable cases where industry practitioners have had success in reducing detrimental strategic behavior including combatting Web spam that attempts to bring questionable content to the top of Web search results.

Sometimes strategic behavior is intended to directly thwart the goal of the designers of the computing system. For example, consider the AdNausium browser extension[112] designed to counteract online advertising by randomly clicking on ads in the background, obstructing the attempt to profile the user, while also imposing costs on advertisers who pay per click. This example also illustrates the sort of value conflict tensions (see Section 2.2) that designers of computing systems must consider, here between consumers and content creators reliant on advertising. Or consider how protesters in Hong Kong (and others) developed new strategies for avoiding detection by facial recognition

---

[111] L. Hu, N. Immorlica, and J. Wortman Vaughan, 2019, "The Disparate Effects of Strategic Manipulation," Pp. 259-268 in *Proceedings of the Conference on Fairness, Accountability, and Transparency,* ACM (Association of Computing Machinery), New York, https://doi.org/10.1145/3287560.3287597; S. Milli, J. Miller, A.D. Dragan, and M. Hardt, 2019, "The Social Cost of Strategic Classification," Pp. 230-239 in *Proceedings of the Conference on Fairness, Accountability, and Transparency*, ACM, https://doi.org/10.1145/3287560.3287576.

[112] Adnauseam, "Homepage," https://adnauseam.io.

cameras.[113] These are cases in which a computing system is designed to extract something from users, and the users resist that extraction.

In other cases, strategic behavior is deployed in order to co-opt the computing system, turning its outputs to the advantage of the user. This co-option can be relatively trivial, as with the way users of Microsoft's Tay taught it swear words and hate speech before Tay was quickly taken down. But this dynamic can also have very significant social and political consequences. For example, content creators have sought to understand the recommender systems that allocate attention online, in order to optimize the visibility of their content. This interplay is described at length in Bucher.[114] Some argue that the tendency of social media companies to promote highly engaging, potentially divisive content has been operationalized by extremists in order to advance their economic and political interests, with deleterious effects on democratic public cultures.[115]

In addition, in recent years computing researchers have had to pay attention not only to strategic behavior by users, but to strategic behavior by competing computing systems, such as generative adversarial networks, whose role is to learn the behavior of a computing system and then confound it. Importantly, this feedback loop between competing systems—one producing increasingly difficult problem instances and one trying to learn from them—has itself led to powerful new methods for generating and classifying image and text data.[116]

Of course, strategic behavior by those who are subject to a computing system need not always be socially deleterious. If the system is optimizing for some feature that is itself a valuable attribute for people to display, then strategic behavior can actually advance the goals of the system.[117] This depends on the system being sufficiently interpretable to those affected by it, so that they can rationally respond to the incentives it creates.[118] Similarly, strategic behavior to resist the use of computing systems to exercise social control, or to extract value from users, should generally be welcomed.

---

[113] A. Holmes, 2019, "These Clothes Use Outlandish Designs to Trick Facial Recognition Software into Thinking You're Not Human," *Business Insider, Australia,* https://www.businessinsider.com.au/clothes-accessories-that-outsmart-facial-recognition-tech-2019-10.

[114] T. Bucher, 2018, *If… Then: Algorithmic Power and Politics,* Oxford Scholarship Online, Oxford University Press, United Kingdom, https://oxford.universitypressscholarship.com/view/10.1093/oso/9780190493028.001.0001/oso-9780190493028.

[115] See, for example, K. Munger and J. Phillips, 2020, "Right-Wing YouTube: A Supply and Demand Perspective," *The International Journal of Press/Politics,* https://doi.org/10.1177%2F1940161220964767; C. Metz, 2021, "Feeding Hate with Video: A Former Alt-Right YouTuber Explains His Methods," *The New York Times,* https://www.nytimes.com/2021/04/15/technology/alt-right-youtube-algorithm.html.

[116] I. Goodfellow, J. Pouget-Abadie, M. Mirza, B. Xu, D. Warde-Farley, S. Ozair, A. Courville, and Y. Bengio, 2020, "Generative Adversarial Networks," *Communications of the ACM* 63(11):139-144.

[117] J. Kleinberg and M. Raghavan, 2019, "How Do Classifiers Induce Agents to Invest Effort Strategically?" Pp. 825-844 in *Proceedings of the 2019 ACM Conference on Economics and Computation,* Association for Computing Machinery Phoenix, AZ.

[118] A.D. Selbst and S. Barocas, 2018, "The Intuitive Appeal of Explainable Machines," *Fordham Law Review,* 87:1085-1139, https://dx.doi.org/10.2139/ssrn.3126971.

The strategic behavior of users and agents in response to computing systems poses a challenge for responsible computing research, because it requires integrating computing research with a deep understanding of human psychology and social scientific insight into the effects of computing systems when deployed in the real world, often at massive scale.

Fulfilling Societal Responsibilities

Disparate access to technologies. Computing technologies have become increasingly essential to economic, social, and political life, and have in some cases dramatically improved economic opportunity for individuals and small businesses through, for example, better access to markets and financial services. Such benefits point to the value to society of making technologies more widely accessible.

At the same time, the consequences of disparities in access to them have grown more pronounced. The disparities have multiple sources including the cost of computing hardware, software, and communications and other services. Although the price-performance ratio for smartphones and laptops has fallen over time, they are still expensive on an absolute basis. Another source of disparity is availability—broadband Internet service is unavailable in some areas, especially rural ones, and even when service is available it may be of poor performance or high cost. Caps on monthly data use are another constraint. Last, not everyone has the skills and experience needed to make effective use of computing technologies.

Several presenters to the committee discussed ways these issues play out in the delivery of health care. As Vickie Mays put it, computing technologies were enormously beneficial during the COVID-19 pandemic but because those technologies are not distributed in an equitable fashion, the benefits of the computing technologies have been unavailable to the very people who need them most.[119] For example, those without broadband Internet were at a distinct disadvantage in obtaining information about the COVID-19 vaccine let alone in making an appointment to receive it. Abi Pitts echoed these concerns, noting that the shift to telemedicine during the pandemic took place without full consideration of the impacts on more vulnerable populations, thereby widening disparities in access to health care in those groups.[120]

For computing researchers, one lesson is that disparities in access remain an issue for any service that needs to reach the entire population. It is also a reminder that computing research aimed at reducing cost, increasing availability, or improving usability

---

[119] V. Mays, presentation to the committee on May 6, 2021, University of California, Los Angeles.
[120] M.A. Pitts, presentation to the committee on May 11, 2021, Stanford University School of Medicine/Santa Clara Valley Medical Center.

can reduce disparities in access to the services needed for well-being and participation in society.

Governance principles for new technologies. When the results of computing research are integrated into systems with potential societal impact, some facets of social responsibility need to be addressed through governance mechanisms and by regulatory bodies. In many of these settings there are diverse goals and incentives and varying regulatory and governance policies and structures. There may also be considerable variation in the maturity of these policies and structures. Institutions across technologies and domains, and institutions responsible for developing or carrying out oversight can have blind spots or inadequate controls and governance mechanisms and sometimes capacity challenges. Some particularly distinct and challenging arenas include the following: commercial sector developing computing technologies for highly competitive and fast-moving markets and innovation arenas; public-sector services such as health, education, and human services; and national security.

A variety of approaches are possible, and a variety of challenges arise. For instance, possibilities for oversight in the commercial sector include internal advisory boards (e.g., Microsoft's Aether Committee, Office of Responsible AI, and Responsible AI Strategy in Engineering) and risk management activities by a company's board of directors. In the public sector, governance procedures are needed to ensure that public goals and incentives inform technology procurement and deployment. For national security, governance and oversight are particularly difficult because of secrecy needs but also ever more important.

The governance challenges in such circumstances are not the direct responsibility of computing researchers but they do create opportunities for computing researchers to engage with government agencies, private-sector institutions, and civil society to help develop or enhance governance principles and frameworks.

3.4 SYSTEM ROBUSTNESS

In computing R&D, some ethical dilemmas and problematic societal impact are difficult to predict. Many of these dilemmas, including some of those described in earlier sections, involve interactions between technology and society that are novel and unprecedented, and so ongoing monitoring and reevaluation (as Chapter 2 argues) and willingness to adapt the technology post-deployment are the only possible ways to adequately handle ethical and societal impact issues. Many other ethical and societal impact problems in computing are not, however, of this nature. Rather, they arise from failures to apply known best practices for ethical design, and failures to devoting enough time, thought and imagination to pondering ways a technological system might be used or exploited in unanticipated manners. This section describes several of the major computing technical arenas in which failures are of this second nature. Recommendation 7 provides several steps researchers can take to avoid them.

3.4.1 Trustworthy, Secure, and Safe Systems

Typical computer scientists' answers to questions of what makes a program "secure" is to provide a specific set of attacks that should be prevented, or a particular collection of defenses that a system should employ. For example, data theft is bad, therefore a system must enforce access controls and keep data encrypted by default; running arbitrary code from unknown origins is bad, so a system must run malware scanners to identify and quarantine such code. Such responses reflect the nebulous nature of the concept of "security." Conceiving of security in terms of the mere composition of known techniques threatens to miss the forest for the trees because systems differ in their purposes, users, and data. Designing a trustworthy system thus cannot be a checklist-based exercise. Instead, the design effort must center around the following idea: A secure system is one that behaves correctly, despite the active malice or unintentional incompetence of users, administrators, and developers.

Frustratingly, this idea, when applied to any *particular* system, raises more questions than it answers. Crisply defining the correct behavior for a non-toy system is hard. Indeed, generating explicit models for what a program should and should not do is a primary challenge of making a trustworthy system. Furthermore, as a system, its users, and the rest of its operating environment changes, the definition of "correct behavior" may change. There are, nonetheless, a variety of well-known techniques and practices that can be brought to bear to ensure that a program behaves correctly.

Many of these approaches have been known for quite a long time. A classic 1975 paper by Saltzer and Schroeder,[121] for example, enumerates design principles that remain instructive today. These include open design (which states that systems should be designed in a way that makes them secure even if attackers know everything about the system design except for the cryptographic keys used by the system), and least privilege (which means that each program and user in a system should have the least amount of authority necessary to perform the relevant tasks).

Experience shows that security must be considered during the earliest phases of system design. This advice can be difficult for computing researchers to follow, though,

---

[121] J.H. Saltzer and M.D. Schroeder, 1975, "The Protection of Information in Computer Systems," *Proceedings of the IEEE* 63:1278-1308, https://doi.org/10.1109/PROC.1975.9939.

because many of the systems that they build are not ostensibly oriented around security goals. However, as computational technology becomes increasingly ubiquitous, reaching deeper into people's lives and operations of the public and private sectors, the consequences of incorrect system behavior are multiplying.

For example, consider Internet of Things (IoT) systems, which embed sensors, actuators, and other computational elements into homes, factories, office buildings, and cityscapes. IoT systems enable "smart" physical environments that can self-adjust their temperature or respond to other environmental stimuli. IoT systems also allow users to perform remote inspection or administration of locales for which placing a human on-site would be difficult, inconvenient, or expensive. Adding technology to an environment that previously did not embed technology, however, exposes that environment to new risks that must be explicitly considered. For instance, embedded medical devices introduce the risk that a person's health becomes directly vulnerable to attacks, and placing IoT technology inside of cars introduces new security threats to cars because attackers can break into your car not only by physically breaking a window but also by hacking into the IoT subsystem and getting the car to unlock its own doors.

Unfortunately, business interests in bringing IoT technology to market and an absence of government regulation led to a flood of insecure systems. For example, smart light bulbs controllable via Wi-Fi networks used weak encryption systems, exposing Wi-Fi passwords to network eavesdroppers.[122] The Mirai botnet exploited the fact that many network-controllable video cameras drew their administrator login credentials from a small set (e.g., username "root" and password "admin1234"); using those credentials, Mirai logged into and commandeered hundreds of thousands of devices, using them to generate hundreds of gigabits per second of denial-of-service traffic.[123] In some cases, fixing these IoT security problems was impossible because the devices stored their software in read-only memory.

Experience also shows that developing a "threat model," the formal or semi-formal description of the security problems that are in-scope and out-of-scope for a system to prevent, is likewise important because it drives subsequent design work: if a threat is in-scope according to the threat model, then the design must handle that threat. When technologists craft an explicit threat model, they are forced to think like attackers and reckon with possible vulnerabilities; this reckoning typically helps technologists to better understand their systems, and to remove or mitigate possible vulnerabilities. Experience also shows the importance of a post-deployment security strategy for discovering,

---

[122] D. Goodwin, 2014, "Crypto Weakness in Smart LED Lightbulbs Exposes Wi-Fi Passwords," *Ars Technica*, https://arstechnica.com/information-technology/2014/07/crypto-weakness-in-smart-led-lightbulbs-exposes-wi-fi-passwords.

[123] M. Antonakakis, T. April, M. Bailey, M. Bernhard, E. Bursztein, J. Cochran, Z. Durumeric, et al., 2017, "Understanding the Mirai Botnet," *Proceedings of USENIX Security*, https://www.usenix.org/conference/usenixsecurity17/technical-sessions/presentation/antonakakis.

prioritizing, and fixing security bugs is critical; modern devices (whether they be medical or otherwise) are extremely complicated, and thus will almost certainly contain security bugs at deployment time. Even if vendors employ best-practice security measures at design time, after devices are released to actual consumers, they need to constantly monitor those devices for evidence of unexpected security problems. Researchers and developers must consider potential security threats, even if, historically speaking, attackers have ignored a particular kind of system, or have been unsuccessful in subverting that system. As Kevin Fu stressed in his remarks to the committee on medical device security, attackers are clever, motivated, and relentless, so a lack of prior successful exploits does not imply that no security problems will emerge in the future.[124]

Many IoT security follies arose from a neglect to apply these and other well-known best practices for security; others reflected technical myopia in IoT's early days.[125] This instance is an example of where some researchers attempted to anticipate future problems but to little avail because the incentives for industry were in a different direction. IoT devices resemble, approximately, traditional network servers that are exposed to potentially malicious clients. Designers and developers of these devices would have done well to apply the decades of wisdom that the technology industry has accumulated about how to protect network servers. This neglect of wisdom from past experiences and measures to avoid or circumvent serious ethical and societal impact problems is true of many other software systems.

Research systems often trod new ground; they are speculative and exploratory, making it hard to predict how intentional evildoing or accidental misbehavior could force an unfinished system to behave incorrectly. To protect these systems, researchers require a knowledge of history and prior art, but perhaps more importantly, some imagination. For example, consider machine learning (ML). Broadly speaking, the goal of an ML system is to analyze a piece of input data and output a classification or prediction involving that data point. To generate such analyses, an ML system must first be trained using a training data set. The quality of that data set influences the quality of the learned observations.

Specific experiences with social networking applications and comment sections teach that a non-trivial number of users will intentionally submit maliciously designed content. The recent emphasis in the ML community on poisoned data sets[126] and

---

[124] K. Fu, presentation to the committee on May 11, 2021, U.S. Food and Drug Administration.

[125] Early on, IoT CPUs were so low-powered that they could not do "real" cryptography or accept the overhead of standard security measures. Some rushed to deploy devices lacking the power to implement proper security measures. Per Moore's Law, that changed, but too many developers never twigged to the change, instead repeating the shibboleths from a few years earlier.

[126] A. Shafahi, W.R. Huang, M. Najibi, O. Suciu, C. Studer, T. Dumitras, and T. Goldstein, 2018, "Poison Frogs! Targeted Clean-Label Poisoning Attacks on Neural Networks," arXiv, https://arxiv.org/abs/1804.00792.

adversarially chosen examples[127] recognizes this problem. It would, however, have been better had researchers anticipated malicious behavior based on an understanding that the systems they were producing were sociotechnical. To design trustworthy systems, one must always assume that inputs are untrusted by default.

Of course, hindsight is always 20/20. Consider Web technology. Could one fault the developers of the unencrypted, unauthenticated HTTP protocol for not predicting that the protocol would become a foundational technology that would serve as a conduit for emails, financial information, and other sensitive data? Could one blame the inventors of Web cookies for not predicting that cookies, intended to make online shopping carts easier to implement, would eventually be used to support a vast ecosystem of online tracking? Perhaps not—the Web has been "catastrophically successful" in a way that would boggle the mind of a computer scientist from the early 1990s. However, the modern era is a different one. Now, and for the foreseeable future, technology will be embedded into every aspect of human life. As a result, computer scientists have a solemn responsibility to ponder what would happen if their technologies became catastrophically successful. To do so, computer scientists must think holistically, at a sociotechnical level, about what their programs should and should not do, and then take explicit steps to enforce those expectations.

3.4.2 Software Engineering: Lessons and Limitations

In computationally intensive fields, code is a frequent research product, with that code being adopted by other researchers and sometimes making its way into products. Bugs have consequences.[128] A range of software engineering best practices have been developed to help ensure the trustworthiness of a program,[129] including the following:

- *Design before implementation:* Describe what the code should do, plan the architecture of the software artifact, and then fill in the algorithm details.
- *Test:* Implement tests of the code's correctness.
- *Peer review:* Have another developer examine the code and provide feedback.
- *Document:* Document the code.

---

[127] I. Goodfellow, J. Shlens, and C. Szegedy, 2015, "Explaining and Harnessing Adversarial Examples," arXiv, https://arxiv.org/abs/1412.6572.

[128] See, e.g., A. Tay, 2020, "Three Ways Researchers Can Avoid Common Programming Bugs," *Nature Index*, https://www.natureindex.com/news-blog/three-ways-researchers-science-can-avoid-common-programming-bugs-errors; D.A.W. Soergel, 2015, "Rampant Software Errors May Undermine Scientific Results," *F1000Research* 3:303, https://doi.org/10.12688/f1000research.5930.2.

[129] A. Trisovic, M.K. Lau, T. Pasquier, and M. Crosas, 2021, "A Large-Scale Study on Research Code Quality and Execution," arXiv, https://arxiv.org/abs/2103.12793.

Computing researchers, with a focus on exploring new ideas and developing new kinds of systems, typically do not follow these practices. When research code is subsequently used by other researchers, its flaws may adversely affect their research and then further leak into the scientific literature. In one notable case, a bug in a script led to potentially incorrect findings in over one hundred publications.[130] Also, given the rapid pace of technological development, research code may easily make its way into products.

Software engineering best practices are also limited in the range of software to which they apply. As the nature of software artifacts evolves, best practices development can struggle to adapt to technological advances. For example, standard software engineering best practices are not adequate for machine learning artifacts (models), which require different kinds of testing.[131] More generally, software engineering best practices do not encourage or reward the consideration of such downstream impacts of the code as unintended consequences and unforeseen uses and misuses. They assume the purpose, scope, and application of a project are already defined, and these assumptions discourage the type of critical, creative, big-picture thinking necessary for responsible computing research artifact release.[132]

3.4.3 Data Cleaning and Provenance Tracking

Data are a central component of certain areas of computing research (e.g., data science, HCI, and much of artificial intelligence), but computing researchers are generally not taught the basics of data management and handling. The Internet and the open science movement have made certain types of data very easy to obtain, most notably data generated by and about people who are on the Internet and research data in publications. This ease with which researchers can find and use "found data" obscures a plethora of concerns critical to responsible computing research, each with potential ethical and societal impact.

In particular, computing researchers undertaking a data-intensive research project should at the start of using any data set ask several questions about the data. The five key questions below use as an example OpenWebText2 (a corpus of user submissions to the social media platform Reddit).[133]

---

[130] J. Bhandari Neupane, R.P. Neupane, Y. Luo, W.Y. Yoshida, R. Sun, and P.G. Williams, 2019, "Characterization of Leptazolines A-D, Polar Oxazolines from the Cyanobacterium *Leptolyngbya sp.*, Reveals a Glitch with the 'Willoughby-Hoye' Scripts for Calculating NMR Chemical Shifts," *Organic Letters* 21(20):8449-8453, https://pubs.acs.org/doi/10.1021/acs.orglett.9b03216.

[131] A. Paleyes, R.-G. Urma, and N.D. Lawrence, 2020, "Challenges in Deploying Machine Learning: A Survey of Case Studies," arXiv, https://arxiv.org/abs/2011.09926.

[132] J. Gogoll, N. Zuber, S. Kacianka, T. Greger, A. Pretschner, and J. Nida-Rümelin, 2021, "Ethics in the Software Development Process: From Codes of Conduct to Ethical Deliberation," *Philosophy and Technology*, https://link.springer.com/article/10.1007/s13347-021-00451-w.

[133] See OpenWebText2, "Welcome!" https://openwebtext2.readthedocs.io/en/latest/?badge=latest.

- *Is this data set fit for purpose?* Is this data set a good fit for the research question being addressed? Because some data are easy to obtain, the temptation may be to use it even if it's not a good fit for the research project. For example, OpenWebText2 would not be appropriate for studying language change over time because it only covers the years after 2005.
- *Is this data set permissible to use?* Data may be available yet still protected. For example, some data from OpenWebText2 may be copyrighted or protected by user agreements. Copyrighted data may still be permissible for use in research under fair use,[134] but it may be harder to defend the use of data protected by user agreements.[135]
- *Does this data set comprise an appropriate sample?* In the age of "big data," it is easy to assume that any sufficiently large quantity of data is a good sample, but it's impossible to know that without examining the data. For example, OpenWebText2 may seem like a "good sample," but is likely to underrepresent content from China and India, two of the world's largest countries by population. That may or may not be acceptable for the research goal.
- *Does this data set need to be protected?* For example, data may contain personally identifiable information. Even if a data set has been cleaned and anonymized, it may be possible for it to be deanonymized.[136]
- *How should data be cleaned and normalized (i.e., structured in a standardized fashion)?* OpenWebText2 contains not just natural language but also code, links, tables, and so on.

These concerns arise even for data sets curated by other researchers. A new use requires reconsideration[137] and despite having a data sheet (as a data sheet may be incomplete, incorrect or out of date[138]), derivatives of a data set may pose additional concerns.[139] For example, the GPT series of models from OpenAI, trained on OpenWebText

---

[134] A. Levendowski, 2017, "How Copyright Law Can Fix Artificial Intelligence's Implicit Bias Problem," *93 Washington Law Review* 579(2018), https://ssrn.com/abstract=3024938.

[135] See ACLU, 2020, "ACLU Sues Clearview AI," https://www.aclu.org/press-releases/aclu-sues-clearview-ai.

[136] A. Narayanan and V. Shmatikov, 2019, "Robust De-Anonymization of Large Sparse Datasets: A Decade Later," https://www.cs.princeton.edu/~arvindn/publications/de-anonymization-retrospective.pdf.

[137] A. Birhane and V.U. Prabhu, 2021, "Large Image Datasets: A Pyrrhic Win for Computer Vision?" Pp. 1536-1546 in *2021 IEEE Winter Conference on Applications of Computer Vision (WACV)*, https://ieeexplore.ieee.org/document/9423393.

[138] E. Yang and M.E. Roberts, 2021, "Censorship of Online Encyclopedias: Implications for NLP Models," Pp. 537-548 in *Proceedings of the ACM Conference on Fairness, Accountability, and Transparency*, https://doi.org/10.1145/3442188.3445916; R. Baeza-Yates, 2018, "Bias on the Web," *Communications of the ACM* 61(6):54-61, https://doi.org/10.1145/3209581.

[139] J. Doge, M. Sap, A. Marasović, W. Agnew, G. Ilharco, D. Groeneveld, M. Mitchell, and M. Gardner, 2021, "Documenting Large Webtext Corpora: A Case Study on the Colossal Clean Crawled Corpus," https://aclanthology.org/2021.emnlp-main.98.pdf.

and other data, are not merely condensed versions of the input data. They can be used to generate text, for example to help student learners of English, or in spam bots. The Copilot model from Microsoft, trained on code from GitHub, can reproduce code that is not licensed for reuse.[140] Furthermore, owing to variations in data collection, sampling, cleaning and normalization, it can be very difficult to trace exactly what data is included in any derivative of a data set.

Last, the use of data from humans creates specific challenges. Institutional review board reviews cover only certain types of data intensive research, and only some facets of the use of data from humans, as discussed in a recent Department of Health and Human Services panel.[141] By contrast, responsible computing research requires looking at a broader set of issues associated with data-intensive computational modeling.[142] This point was emphasized by Eric Horvitz in his presentation to the committee.[143]

3.4.4 Designing for Responsibility

The usability of a computing technology is a crucial determiner of whether it will function as intended. Its usability, and more specifically its accessibility, will determine the variety of users who will be able to interact with it. Research in HCI has developed various methods and design principles for making computing systems usable and useful among other considerations.[144] These HCI methods can be used to guide not only the design of HCI systems (aka "user interfaces"), but also usability evaluation and often requirements analysis.

More broadly, the theory and methods of design have yielded various tools for imagining possible designs and the futures they might engender. For responsible computing research these tools enable the design of computing methods and systems that increase societal good as well as ones that avoid or mitigate unforeseen consequences and unintended uses of computing research outcomes.[145] Research about values in

---

[140] GitHub Gist, 2021, "Risk Assessment of GitHub Copilot," https://gist.github.com/0xabad1dea/be18e11beb2e12433d93475d72016902.

[141] Department of Health and Human Services, "Secretary's Advisory Committee on Human Research Protections (SACHRP) July 21–22, 2021, Meeting," https://www.regulations.gov/docket/HHS-OPHS-2021-0015/document.

[142] S. Santy, A. Rani, and M. Choudhury, 2021, "Use of Formal Ethical Reviews in NLP Literature: Historical Trends and Current Practices," arXiv preprint, https://doi.org/10.48550/arXiv.2106.01105; S.R. Jordan, 2019, "Designing Artificial Intelligence Review Boards: Creating Risk Metrics for Review of AI," Pp. 1-7 in *2019 IEEE International Symposium on Technology and Society (ISTAS)*, https://ieeexplore.ieee.org/document/8937942.

[143] E. Horvitz, presentation to the committee on June 10, 2021, Microsoft Research.

[144] B. Shneiderman and C. Plaisant, 2016, *Designing the User Interface: Strategies for Effective Human–Computer Interaction*, Pearson Education, New York.

[145] J. Van den Hoven, 2013, "Value Sensitive Design and Responsible Innovation," Pp. 75-83 in *Responsible Innovation* (R. Owen, J. Bessant, and M. Heintz, eds.), John Wiley & Sons, Hoboken, NJ; J. Van den Hoven, G.-J. Lokhorst, and I. Van de Poel, 2012, "Engineering and the Problem of Moral Overload," *Science and Engineering Ethics* 18(1):143-155, https://doi.org/10.1007/s11948-011-9277-z.

design[146] provides methods for value sensitive design and for design justice. Participatory design approaches emphasize the active involvement of current or potential users of a system in design and decision-making. Design has become an essential part of HCI and in considering how people really use computer systems as well as an essential complement to ethics and the sociotechnical perspective outlined in Chapter 2.

Early usability efforts focused on error rate, efficiency, learnability, memorability, and satisfaction. Basic usability requires that developers observe the cognitive limits of users. For example, people can remember only roughly 7 ± 2 chunks of information (Miller's Law[147]), limiting their capability to understand dense information on a display. Another constraint is that people can handle only a small number of notifications, often leading to users ignoring alert boxes when the notifications occur too fast, which can lead to safety issues. These cognitive requirements can be most easily viewed in the U.S. General Services Administration's usability guidelines;[148] these should be followed by any system or application aimed at users.

These basic cognitive limitations are too narrow to incorporate the full range of use, however. Understanding that technical systems are inherently sociotechnical systems, and therefore wrapped up in their social context of use, requires additional considerations:

- Designs need to take into account the different kinds of people that may use the system and the contexts of their use. Different user interfaces may be required by different groups of users, for example. Different user groups—such as inexperienced users, expert users (e.g., airplane pilots), elderly, and people with disabilities—may have different usability requirements, such as changes to the user interface or the system functionality. Linguistic variation owing to such factors as accent, dialect, age and gender, or code switching is another important factor for any systems that interact through spoken or written language. Accessibility is still a significant issue for those with disabilities,[149] such as people who are vision impaired or hearing impaired, or those with movement disabilities. System designers should take careful account of the range of accessibility issues. Special guidelines can be found at Web Content Accessibility Guidelines.[150]

---

[146] B. Friedman and D.G. Hendry, 2019, *Value Sensitive Design: Shaping Technology with Moral Imagination*, MIT Press, Cambridge, MA.

[147] G.A. Miller, 1956, "The Magical Number Seven, Plus or Minus Two: Some Limits on Our Capacity for Processing Information," *Psychological Review* 63(2):81-97, https://psycnet.apa.org/doi/10.1037/h0043158.

[148] See U.S. General Services Administration, "Improving the User Experience," https://www.usability.gov.

[149] K. Holtzblatt and H. Beyer, 1997, *Contextual Design: Defining Customer-Centered Systems*, Morgan Kaufmann Publishers, San Francisco, CA.

[150] W3C Web Accessibility Initiative (WAI), "Web Content Accessibility Guidelines (WCAG)," https://www.w3.org/WAI/standards-guidelines/wcag.

- It is also vital to understand how a system might be potentially used in context, for example, in the specifics of social or organizational processes or the constraints of specific classes of users. For example, an application to help homeless people remember their medication might also require considering where they might find refrigeration to store their insulin. Or notifying family members of an elderly person's fall could be seen either as promoting safety or alternatively as invading privacy; developing health applications depends heavily on the specifics of the social context. Understanding the users in their contexts, in study after study, has been seen to facilitate understanding what system capabilities were required for adoption and effective use by differing users.
- One must consider differences in users' mental models (hence requiring training or assistance) as well as their having potentially very different goals.[151] Designers must consider appropriate reward systems or incentive structures in systems that will incorporate groups of users, especially large-scale systems such as social computing or social media systems. Differences in mental models, goals, and reward systems can lead to the benefits and pathologies of anonymity, maladaptive or antagonistic sharing of information, and support for informal roles in organizations and social groupings. Recent research has included a stronger understanding of both network structures, social and computational, including the propagation of information cascades (e.g., misinformation)[152] and the role of networks in sharing expertise.[153]
- Recent work in HCI has included a further reconsideration of usability in the sociotechnical context of use. Computing technologies no longer replace current work practices or standard operating procedures with digital practices; much of people's everyday lives have already become digital. HCI is now envisioning how to design new computational contexts and how to include all types of users into those contexts.

These requirements for usable systems are not easily considered. Important methods for uncovering the requirements for research projects and products include task analyses and cognitive walkthroughs.[154] Testing usability can include think-aloud evalu-

---

[151] W.J. Orlikowski, 1995, "Learning from Notes: Organizational Issues in Groupware Implementation," Pp. 197-204 in *Readings in Human-Computer Interaction*, Morgan Kaufmann Publishers, San Francisco, CA.

[152] D. Easley and J. Kleinberg, 2010, *Networks, Crowds, and Markets*, Cambridge University Press, United Kingdom.

[153] J. Zhang, M.S. Ackerman, and L. Adamic, 2007, "Expertise Networks in Online Communities: Structure and Algorithms," Pp. 221-230 in *Proceedings of the 16th International Conference on World Wide Web*, https://www2007.org/papers/paper516.pdf.

[154] B. Shneiderman and C. Plaisant, 2016, *Designing the User Interface: Strategies for Effective Human-Computer Interaction*, Pearson Education, New York.

ations and many other techniques.[155] As researchers and practice began to consider system use in its social context, additional methods were developed, including contextual inquiry.[156] General guidelines for developing usable systems have been developed,[157,158] and software developers would be remiss if they do not use all methods applicable to their systems.

3.5 LIMITS OF A PURELY COMPUTING-TECHNICAL APPROACH

The sections below discuss two areas of concern—privacy and content moderation—that have arisen with the proliferation of highly networked computing environments and data-dependent AI systems in widespread use. They illustrate the need for close integration of a wide range of disciplinary perspectives pertaining to ethical and societal factors of the type described in previous sections—including social and behavioral science, policy, and governance—and the importance of bringing these perspectives into consideration at the earliest stages of the computing research pipeline. They also show the limits on what can be achieved from a purely technical point of view if this range of perspectives is not invoked until after systems have been deployed. As part of this integration, the computing research community has important roles to play in informing approaches to the key questions in these domains and in developing new methods to assist in addressing them.

3.5.1 Limits to Privacy Protection and Risk Assessments

Computational systems regularly process and store sensitive information. Thus, privacy is a central design principle across many areas of computing. However, the mere act of even defining privacy is a difficult one because there are different meanings in different contexts (see also Section 2.1, "The Value and Scope of Ethics"). For example, viewed through the narrow lens of access control and system administration, privacy might refer to who may control how your data are accessed, or who can associate their data with yours, or who can influence the decisions that you make within the system. However, privacy also speaks to higher-level concepts involving human expression and societal organization; privacy rules help to define the relationships between individuals and institutions (public and private) with differing amount of power, encouraging or discouraging

---

[155] Ibid.
[156] K. Holtzblatt and H. Beyer, 1997, *Contextual Design: Defining Customer-Centered Systems,* Morgan Kaufmann Publishers, San Francisco, CA.
[157] U.S. General Services Administration, "Homepage," https://www.usability.gov.
[158] J. Nielsen, 1994, *Usability Engineering,* Morgan Kaufmann, San Francisco, CA.

individuals to engage in free expression, association, and intellectual engagement without the chilling effects of being monitored or restricted.

All of these concerns long predate the development of computing systems. However, the rise of computing has added new urgency for reasons of automation and scale. Data sets that were public in theory but of limited accessibility in practice are no longer costly or difficult to obtain; for example, information about home sales and political donations are now digitized and straightforward for almost anyone to download and analyze. Although the democratization of access has many positive aspects, it also raises new privacy questions.

Furthermore, as various aspects of life increasingly involve an online component, new data sets are being generated and being made widely available. These data sets, sometimes held privately by a single company, other times shared with other businesses via often opaque arrangements, have dramatically increased the amount of data available to analyze. The rise of cheap, commoditized computing-as-a-service has lowered barriers to storing that information and extracting insights from it.

Given all of this, thinking about privacy in computational settings demands a reckoning with several fundamental questions. First, what kinds of privacy approaches are desirable? Second, what kinds of privacy approaches are technologically possible? Third, how can different approaches to privacy protection in different countries or regions best be reconciled or accommodated? The questions are related; pondering them together is helpful for identifying both risks and opportunities. For example, imagine a database that stores sensitive information about a variety of users. A desirable privacy goal might be "queries about this database do not reveal whether my particular information resides in the database." Starting from that high-level goal, a variety of computer science research has explored the technical possibility of achieving it, using approaches like differential privacy[159] (which itself was motivated by a desire to minimize the information leakage allowed by earlier anonymization techniques).[160] As information-hiding techniques become more advanced (i.e., as the scope of what is technologically possible becomes broader), the notion of which privacy approaches are desirable will change.

Of course, no solution is perfect. For example, differential privacy is not effective if the size of the aggregate data set is insufficient to mask the contribution of specific individuals. Differential privacy also may not provide mechanisms to hide a user's IP address from the differentially private database server. Issues like these often arise when

---

[159] C. Dwork and A. Roth, 2014, "The Algorithmic Foundations of Differential Privacy." *Foundations and Trends in Theoretical Computer Science* 9(3-4):211-407, http://dx.doi.org/10.1561/0400000042.

[160] P. Ohm, 2009, "Broken Promises of Privacy: Responding to the Surprising Failure of Anonymization," *UCLA Law Review* 57:1701, 2010, University of Colorado Law Legal Studies Research Paper No. 9-12, https://ssrn.com/abstract=1450006; L. Sweeney, 1997, "Weaving Technology and Policy Together to Maintain Confidentiality," *The Journal of Law, Medicine, and Ethics* 25(2-3):98-110, https://journals.sagepub.com/doi/10.1111/j.1748-720X.1997.tb01885.x.

considering a third fundamental question: what are the privacy implications of how an *end-to-end* system is designed? For example, once network communication between clients and servers is introduced, privacy of the network data and client-specific information like IP addresses and browser fingerprints must be considered. Access control is also important—how does a system determine whether a user is authorized to access a particular piece of sensitive data?

Various point solutions for each of those challenges exist. For example, encryption provides data confidentiality, while mixnets[161] can hide IP addresses by routing traffic through intermediate servers. However, integrating specific technical solutions or research ideas into a cohesive system is challenging. For example, Tor[162] uses both mixnets (to hide client IP addresses) and encryption (to hide the identity and content of visited websites). However, a network observer can still confirm that a user has visited given sites by looking at the size and arrival times of network packets, even though those packets are encrypted.[163] The existence of such side channels is an example of the practical complications that arise when designing privacy-preserving systems.

Tor's designers were aware of side channels involving packet size and packet arrival times. In considering the trade-off between user-perceived performance and expensive techniques for side channel mitigation—a kind of design-values choice that is common for secure systems—they chose performance. This choice was driven in part by an assumption that no realistic network attacker could possess a sufficiently large number of vantage points to run the necessary correlation analyses. At the time of the writing of this report, the number of Tor routers is, however, small enough to make it possible for a nation-state-level actor to subvert or spy on a significant fraction of Tor traffic.

Larger societal concerns add significantly to the complexity of these questions. To the extent that privacy can be conceptualized as a set of conditions on appropriate information flows between different parties,[164] the circumstances that give rise to privacy considerations in everyday interaction are essentially ubiquitous. And as mentioned earlier, many computational settings where privacy concerns arise—with respect to governments, companies, employers, or families—also involve power differentials, and the desire to limit the uses of power. Understanding the real-life relationships between these parties is critical for understanding the desirable technology-mediated privacy

---

[161] D. Chaum, 1981, "Untraceable Electronic Mail, Return Addresses, and Digital Pseudonyms," *Communications of the ACM* 24(2):84-90, https://doi.org/10.1145/358549.358563.

[162] R. Dingledine, N. Mathewson, and P. Syverson, 2004, "Tor: The Second-Generation Onion Router," *Proceedings of the 13th USENIX Security Symposium,* http://usenix.org/publications/library/proceedings/sec04/tech/full_papers/dingledine/dingledine.pdf.

[163] V. Rimmer, D. Preuveneers, M. Juarez, T. Van Goethem, and W. Joosen, 2018, "Automated Website Fingerprinting Through Deep Learning," Network and Distributed Systems Security (NDSS) Symposium 2018, http://dx.doi.org/10.14722/ndss.2018.23115.

[164] H. Nissenbaum, 2011, "A Contextual Approach to Privacy Online," *Daedalus* 140(4):32-48, https://doi.org/10.1162/DAED_a_00113.

relationships between those parties. For example, how is private phone data accessed in the context of intimate partner violence?[165] To what extent should employers be able to surveil their employees for performance or safety reasons?[166] These questions cannot be answered from a purely technical perspective, and significant responsibility rests with companies and governments.

3.5.2 Limits of Content Moderation

The World Wide Web was created to be a platform for communication and expression, and it implements a type of many-to-many interaction that is hard to achieve with other media. Relative to earlier print and broadcast media, it has dramatically lowered the barriers to mass dissemination of content; on the Web, the fact that a piece of content has high production values and wide reach does not necessarily mean that the author had access to expensive mechanisms for creating it. As a result, the Web has led to an abundance of content, both in raw volume and in the heterogeneity of the participants involved in contributing to it.

The complexity of the resulting content creation ecosystem has led to profound challenges for the designers of online platforms. A first challenge lies in determining what kinds of content should be prohibited from a platform. This brings into play classical questions about the fundamental trade-offs involved in balancing the benefits and costs of unrestricted speech—including the representation of diverse viewpoints, and the creation of a public environment where these different viewpoints can engage with and compete with each other—albeit in an environment where these decisions are being made by private operators of online platforms.

Moreover, the problem of managing online content is much broader than simply the question of what should be restricted; platforms must constantly deal with the questions of where users' attention should be directed, and which pieces of content should be amplified. The central role of user attention in these questions was identified early in the history of computing research; for example, Herb Simon wrote in 1971, "In an information-rich world, the wealth of information means a dearth of something else: a scarcity of whatever it is that information consumes. What information consumes is rather obvious: it consumes the attention of its recipients."[167] Editors, publishers, and other gatekeepers have always made decisions about how to allocate attention; however, the

---

[165] D. Freed, S. Havron, E. Tseng, A. Gallardo, R. Chatterjee, T. Ristenpart, and N. Dell, 2019, " 'Is My Phone Hacked?' Analyzing Clinical Computer Security Interventions with Survivors of Intimate Partner Violence," *Proceedings of the ACM on Human-Computer Interaction* 3(CSCW):1-24, https://doi.org/10.1145/3359304.

[166] I. Ajunwa, K. Crawford, and J. Schultz, 2017, "Limitless Worker Surveillance," *California Law Review* 105(3), https://www.californialawreview.org/print/3-limitless-worker-surveillance.

[167] H.A. Simon, 1971, "Designing Organizations for an Information-Rich World," Pp. 37-72 in *Computers, Communications, and the Public Interest* (M. Greenberger, ed.), Johns Hopkins Press, Baltimore, MD; Zeynep Tufekci has made the point that modern censorship takes advantage of this property, by flooding people with information of dubious quality. (Z. Tufekci, 2017, *Twitter and Teargas*, Yale University Press, New Haven, CT.)

challenge of doing so has increased in proportion to the volume of content now available for recommendation, as well as the size of the audience affected; and the concentration of online attention on a few platforms has led to a corresponding concentration of power in the hands of the people running those platforms. Any system design will lead to a distribution of attention that is focused on certain items, viewpoints, and perspectives at the expense of others; this is inevitable in an environment where the abundance of content dramatically outpaces the available attention.

These interlocking questions of amplification and restriction lead to important problems for the computing research community. Even with agreement on general principles for the allocation of user attention, there is only limited understanding of how ranking algorithms and user interface design serve to create the social feedback loops that drive user attention.[168] For content moderation, though there may be agreement that platforms are not simply pipes through which information flows, notably, unlike with privacy, there is not such agreement on what kinds of information should or should not be allowed to spread. That is, there is no consensus on the values and associated goals that content moderation is to realize.

Furthermore, there are significant computational limitations. Determining whether any particular information should *not* be spread is a context-sensitive issue: in isolation a single word or video clip may be fine (e.g., a health care setting in which an anatomical part is mentioned; a sci-fi movie in which a terrorist attack happens), while in others they would not be (e.g., pornography, or a plan to launch an attack). As a result, even where there might be agreement on values and general principles on the types of content that should be restricted, the richness of language and visual media create strong limitations on the power of algorithms to identify particular instances of prohibited content.[169] In addition, people labeling data do not always agree on whether a particular piece of content should be blocked, so it is difficult to get good quality control for training data. Last, languages evolve and expressions that were once acceptable may become unacceptable.[170]

One challenge is developing frameworks for considering the types of outcomes to aim to achieve in such environments, and how to achieve them given the strengths and limitations of content moderation and filtering algorithms. Within this broad category of challenges are particular questions concerning social feedback effects in online platforms

---

[168] M.J. Salganik, P. Sheridan Dodds, and D.J. Watts, 2006, "Experimental Study of Inequality and Unpredictability in an Artificial Cultural Market," *Science* 311(5762):854-856, https://doi.org/10.1126/science.1121066.

[169] W. Warner and J. Hirschberg, 2012, "Detecting Hate Speech on the World Wide Web," Pp. 19-26 in *Proceedings of the Second Workshop on Language in Social Media*, https://aclanthology.org/W12-2103.pdf.

[170] Many people think the impressive ability of algorithms to detect spam suggests it would also be possible to detect undesirable content. The detection of spam is not, however, done by natural language processing analysis of what is said in a message, but rather by looking at range of signals (e.g., metadata on the messages, for instance that indicates the source), and on the ability to determine the ways in which money flows to spammers and shut them down.

that are the focus of active research sub-communities. The effect of personalized content filtering on polarization, through the creation of "filter bubbles" and the facilitation of online organizing is one of these central questions;[171] another is the management of misinformation and its effects.[172] The evolution of both of these topics illustrates the inherent challenges in connecting design choices and immediate user responses to long-range outcomes on the platform. This problem of reasoning about long-range outcomes for platform content from the interaction of users and algorithmic filtering and recommendation is the subject of a growing line of research.[173]

An additional issue is that it is very difficult for academic researchers to conduct open scientific research content moderation because of the large amount of data needed and the high risks to those people whose data might be shared in making such data widely available. Greater progress might be made on these topics through more robust connections between industrial and academic research. There are clear structural challenges in creating these connections, including the proprietary interests of the platforms and the privacy interests of their users. But there is progress in exploring mechanisms for these kinds of collaborations,[174] and they may prove crucial for deeper research into design choices for online content and its longer-range societal implications.

---

[171] E. Bakshy, S. Messing, and L.A. Adamic, 2015, "Exposure to Ideologically Diverse News and Opinion on Facebook," *Science* 348(6239):1130-1132, https://doi.org/10.1126/science.aaa1160; D. Freelon, A. Marwick, and D. Kreiss, 2020, "False Equivalencies: Online Activism from Left to Right," *Science* 369(6508):1197-1201, https://doi.org/10.1126/science.abb2428; E. Pariser, 2011, *The Filter Bubble: How the New Personalized Web Is Changing What We Read and How We Think*, Penguin Press, London, England; C. Sunstein, 2014, "The Daily We: Is the Internet Really a Blessing for Democracy?" *Boston Review* 26(3):4, https://bostonreview.net/forum/cass-sunstein-internet-bad-democracy.

[172] C. Wardle and H. Derakhshan, 2017, *Information Disorder: Toward an Interdisciplinary Framework for Research and Policy Making*, Council of Europe report DGI(2017)09, https://rm.coe.int/information-disorder-toward-an-interdisciplinary-framework-for-researc/168076277c.

[173] S. Dean, S. Rich, and B. Recht, 2020, "Recommendations and User Agency: The Reachability of Collaboratively-Filtered Information," Pp. 436-445 in *Proceedings of the 2020 Conference on Fairness, Accountability, and Transparency*, https://doi.org/10.1145/3351095.3372866; M. Mladenov, E. Creager, O. Ben-Porat, K. Swersky, R. Zemel, and C. Boutilier, 2020, "Optimizing Long-Term Social Welfare in Recommender Systems: A Constrained Matching Approach," Pp. 6987-6998 in *International Conference on Machine Learning*, Vienna, Austria, PMLR 119, https://www.cs.toronto.edu/~cebly/Papers/Matching_for_FairRecSys_ICML2020_preconference_version.pdf.

[174] G. King and N. Persily, 2020, "A New Model for Industry–Academic Partnerships," *PS: Political Science and Politics* 53(4):703-709, https://doi.org/10.1017/S1049096519001021.

4

Conclusions and Recommendations

The technologies that computing research yields have transformed every sector of the economy, enhanced economic welfare through reduced costs and enhanced productivity, and improved our quality of life in many ways. As computing technology has become more pervasive, however, concerns have mounted that some technologies and uses have had harmful effects on individuals, groups of individuals, and society at large, leading to calls for greater attention to ethical and societal considerations in computing research. (The term "computing research" is used in this report to include research in computer science and engineering, information science, and related fields.)

Chapter 2 of this report describes a set of core ethical concepts (Section 2.1) and fundamental ideas from social and behavioral sciences (Section 2.2) for determining ways those ethical concepts play out in everyday life. Chapter 3 describes a variety of generators of ethical and societal challenges that arise in computing research itself and in the further development and integration of computing research outcomes into deployed technologies. As the discussions in its subsections illustrate, some subfields of computer science and engineering—such as artificial intelligence (AI) and machine learning, computer security, and human–computer interaction—have encountered these issues earlier than others, and their experiences are instructive for the entire computing research community.

This chapter recommends initial practical steps toward ensuring that ethical and societal consequences of computing research are more fully considered and addressed and its obligations to support human flourishing, thriving societies, and a healthy planet are fulfilled. Its recommendations assign responsibilities to actors across the full spectrum of the computing research ecosystem: computer researchers, the computing research community, the scientific and professional societies in which they participate, other scholarly publishers, the public- and private-sector agencies and organizations that sponsor computing research, and the public- and private-sector institutions in which computing research is performed. These practical steps do not relieve public and private organizations from considering other ways to take account of ethical and societal consequences.

Importantly, the recommendations are aimed not just at academic researchers and government research funders even though this study was sponsored by the National Science Foundation (NSF), and the committee could easily have focused its recommendations on NSF and the other federal agencies that sponsor computing research. Industry research is an important contributor to the overall information technology innovation ecosystem, and that ecosystem relies heavily on the exchanges of ideas and researchers between academia and industry. It quickly became apparent to the committee that measures aimed at making government-sponsored academic research more responsible could also be applied in industry settings and that their adoption might well foster better outcomes for the research-performing companies and the overall computing research enterprise as well as fostering societally desirable outcomes.

Accordingly, and with the recognition that the management and shareholders of each company must make their own decisions about what computing research they conduct and how they conduct it, this report adopts inclusive terminology in the recommendations that follow for the actors that sponsor and carry out research that encompasses academic, industrial, and government research organizations. This report adopts the term "researchers" as inclusive of researchers in academia, private-sector and government research institutes, and industry, and uses "research institutions" to encompass colleges, universities, and private- and public-sector research organizations.

Although the incentives of academic and other not-for-profit research institutions differ from those of industry, the recommendations are relevant to all these research environments. It is in the setting of incentives for adopting the recommendations and monitoring their use and outcomes that these types of institutional settings differ.

The term "research community" or "computing research community" is used with respect to issues or recommendations that may not necessarily apply to each individual researcher but that do apply to the community as a whole. The term "research sponsors" is used to include those funding computing research in academia, private-sector and government research institutions, and industry. The term "proposal" includes requests for computing research project funding from any source including government, academic institutions, private philanthropy, and industry and both external and internal support.

The recommendations vary in the particular actors in the computing research ecosystem, individually and in combination, to which they assign particular obligations. Recommendation 1 assigns to researchers and the research community the reshaping of computing research to encompass also the ethics and social and behavioral science expertise needed for responsible computing research. Recommendation 2 calls on research sponsors and research institutions to support the research community in broadening this scope and in defining new kinds of projects and research partnerships to carry out this broader program. Recommendation 3 focuses on education, indicating the need for academic institutions to reshape their curricula in various ways and for scientific and professional societies as well as research sponsors to provide training for computing researchers in practices needed for responsible computing research, both carrying it out and evaluating it. Recommendation 4 complements Recommendation 3 by identifying ways computing research institutions along with research sponsors and scientific and professional societies can provide computing researchers with access to scholars and scholarship in ethics and social and behavioral sciences. Recommendations 5 and 6 focus on the two key actors in the computing ecosystem who vet computing research and can assess whether particular efforts adequately address ethical and societal impacts, namely research sponsors and scholarly publishers. Recommendation 7 addresses computing researchers who develop systems, focusing on the need for them to follow best practices. Recommendation 8 asks all actors in the computing research ecosystem to work together to support better public understanding of computing research and its outcomes.

There is as yet little if any empirical data on the effectiveness of these recommended interventions to draw on. Thus, these recommendations were formulated primarily by considering the actors and leverage points in the computing research ecosystem, drawing on the expertise among the study committee members and presenters to the committee, and reviewing some promising early efforts. As with all innovation in science and engineering, the innovations called for in these recommendations require ongoing assessment and revision to determine what works best—see in particular Recommendations 3.5 (build the capacity to evaluate different approaches for researchers) and 5.4 (evaluate the effectiveness of this report's recommendations for research sponsorship) below.

The subsection "Advancing Diversity, Equity, and Inclusion" in Chapter 3 discusses the importance to responsible computing research of considering diversity, equity, and inclusion (DEI). This report considers these issues throughout many of its recommendations, which it does rather than offering separate recommendations because the need to pay attention to them permeates the challenge of responsibility in computing research. In particular, these issues are reflected in recommendations addressing the potential negative impacts of different technologies on underrepresented groups, which is a DEI consideration, and those addressing the value of diversifying the set of stakeholders who inform technological design choices. Much work has been done in the scientific community on DEI challenges in general by other groups with expertise across DEI areas. For example, the National Academies of Sciences, Engineering, and Medicine have compiled a set of reports on diversity and inclusion in science, technology, education, mathematics, and medicine.[1] These concerns are important, and the committee urges all parties engaged in computing research endeavors to act on the recommendations from these reports and those of other groups.

The recommendations provide practical ways for computing researchers and the computing research community to address the situations, conditions, and computing practices discussed in Chapter 3 that have potential to raise ethical challenges and societal concerns. Of special note is the need to address these challenges and concerns at the societal level and not just the individual level. In particular, researchers have an ethical responsibility to consider that computing research can generate or exacerbate not only risks to autonomy, well-being, privacy, and other individual-level intrinsic and instrumental ethical values, but also extreme risks of safety and security breakdowns that could physically harm large numbers of people and society more generally.

Two conclusions from the analyses in Chapters 2 and 3 underlie the recommendations that follow. These conclusions make clear that computing research needs fundamentally to broaden the scope and set of issues it needs to take into account. There may be costs associated with the changes called for in the recommendations below, but they are costs necessary for achieving responsible computing research.

> **Conclusion 1.** To be responsible, computing research needs to expand to include consideration of ethical and societal impact concerns and determination of effective ways to address them.
>
> **Conclusion 2.** To be responsible, computing research needs to engage the full spectrum of stakeholders and deploy rigorous methodologies and frameworks that have proven effective for identifying the complicated social dynamics that are relevant to these ethical and societal impact concerns.

In keeping with the study's Statement of Task, the report's recommendations also do not directly address government regulation of computing technologies including corporate computing research. However, they do discuss ways that the computing research

---

[1] See National Academies of Sciences, Engineering, and Medicine, "Diversity and Inclusion in STEMM Collection," https://www.nap.edu/collection/81/diversity-and-inclusion-in-stemm.

community can help inform government action in this space. As discussed in Chapter 1, the design and deployment of computing technologies that are raising societal and ethical concerns are shaped by a combination of corporate decision-making, incentives set by the market and government regulation, and decisions made by organizations in acquiring the technologies. These factors are the proper realm of societies, who determine norms, and of governments, which institute mechanisms to enforce those norms. Nevertheless, computing researchers have responsibilities related to such societal and ethical concerns. These responsibilities include fully disclosing the capabilities and limitations of their research results and advising the public and governments on areas where adverse impacts may occur and government regulation or changes to corporate governance may be needed. This analysis supports a third conclusion that underlies the recommendations below:

> **Conclusion 3.** For computing technologies to be used responsibly, governments need to establish policies and regulations to protect against adverse ethical and societal impacts. Computing researchers can assist by revealing limitations of their research results and identifying possible adverse impacts and needs for government intervention.

The recommendations are listed in a logical order, not by priority. They differ in the resources, time, and energy they will require. They are intended to work together to enable the computing research community to conduct responsible research. Some of the recommendations are aimed at proactively promoting good while others aim to mitigate or minimize potential harms.

They aim to help the computing research community be more proactive in anticipating and avoiding potential harms rather than, as is generally the case today, only reacting when bad things happen. By taking these steps at the research stage, two kinds of important downstream effects are possible. First, as responsible computing research is taken up by other researchers and technology developers and deployers, it will serve as a model for them to be responsible as well. Second, the recommendations for changes in computing education will help ensure that future computing professionals across industry, including those in product groups and leadership and governance positions, not just research, are better equipped to address ethical and societal concerns.

As noted in the introduction, the recommendations speak to all computing researchers. The recommendations also speak to those whose scholarship and expertise is in disciplines that have studied moral reasoning and those whose scholarship examines the place of science and technology in the world—particularly in philosophical ethics and the theory of sociotechnical systems. The multidisciplinary efforts recommended will involve much more than straightforwardly applying existing theory. They will require deep engagement of such scholars and computing researchers (Chapter 2).

4.1 RESHAPE COMPUTING RESEARCH

Recommendation 1. The computing research community should reshape the ways computing research is formulated and undertaken to ensure that ethical and societal consequences are considered and addressed appropriately from the start.

Just as the cost of addressing problems after the fact is often much higher than the cost of addressing them during the initial design of a system, so too in research it is much easier to address oversights, unanticipated consequences, and unexpected outcomes if more thought is given to these issues at the outset. (An analogous claim has been made with respect to computer security: that it must be considered at the outset rather than bolted on later.) To promote more positive outcomes and to avoid, mitigate, or otherwise address the potential negative consequences of computing research, computing researchers need to draw on expertise from a variety of domains including those in the humanities and social and behavioral sciences; integrate into their research plans engagement with the various populations who are affected by the outcomes of the research (e.g., conclusions, predictive models, or artifacts); and be transparent about the limitations of such outcomes. This reshaping must be an ongoing process with repeated engagements with experts in a variety of domains.

Recommendation 1.1. Research projects should incorporate applicable and relevant expertise in social and behavioral sciences, ethics, and any domains of application the project includes.

The humanities and social and behavioral sciences provide methods and intellectual approaches that are relevant to anticipating and understanding the effects that technologies are likely to have on people, institutions, and society. Researchers in other fields have expertise that can help computing researchers to determine ways their research can have the impact they desire. To incorporate such knowledge, computing researchers should draw on such expertise. For this collaborative endeavor to succeed, computing research and scholars with expertise in these other fields must each acquire sufficient familiarity with each other's approaches and methods to have meaningful conversations.

Such familiarity can be acquired through either course work or independent study. Recommendation 3.1 addresses steps that universities can take. For research projects aimed at applications in particular domains (e.g., health care or education), it is likewise crucial to have the participation of experts knowledgeable about those domains.

> **Recommendation 1.2. Projects that include among their aims the achievement of societally relevant outcomes should engage stakeholders from the start of design through deployment, testing, and redesign processes, and employ design teams that include domain experts and social and behavioral scientists to help ensure that the project is solving the right problems.**

If they do not engage the relevant stakeholders, computing researchers run the risk of building systems that may work for themselves (or their friends and colleagues) but may not perform equally well for other populations including ones not originally targeted by the developer. Put another way, builders of computing technology cannot approach design in an abstract manner but must consider the context of where and how the system will be used, who will use the system, and the risks and benefits to different groups of both users and others who will be affected by it. It is important for such projects also to include expertise in the well-established methods, such as participatory design, for accounting for these diverse factors.

> **Recommendation 1.3. Research projects that produce artifacts or research results likely to be adopted in other research should consider and report on possible limitations such as potential biases or risks of applying them to other problems or in other contexts.**

Many research projects produce artifacts, algorithms, or other methods that have the potential to be widely used by other researchers or by industry. Examples include open-source software, data sets for machine learning, and models (including models such as large language models, which are trained on sufficiently broad data and of a sufficient scale that they are adaptable to a wide range of tasks). Release of such artifacts magnifies their impact and fosters reproducibility yet release of very early research outputs, including code, data sets, and tools, poses potential risks. Any research output can potentially be misused or applied inappropriately, in a context that differs significantly from the one envisioned by the original researchers. Potential forms of documentation include warning statements, user guides, data sheets, model cards, use parameter specifications, and recommended evaluation metrics.

4.2 FOSTER AND FACILITATE RESPONSIBLE COMPUTING RESEARCH

Recommendation 2. The computing research community should initiate projects that foster responsible computing research, including research that leads to societal benefits and ethical societal impact and research that helps avoid or mitigate negative outcomes and harms. Both research sponsors and research institutions should encourage and support the pursuit of such projects.

Responsible computing research involves both efforts to mitigate potential harms arising from the use of computing research (preventive) and research activities that have direct positive social good (proactive) with consideration to the scale and impact of those potential impacts. The multidisciplinary nature of both types of work means that they are often harder to fund; both the disciplines and the majority of funding opportunities are siloed. Both types also involve the engagement of experts in the domain(s) of use, which requires additional investment.

Recommendation 2.1. Research sponsors should develop programs aimed at approaches and tools for reducing or mitigating societally harmful characteristics of computing technologies.

Examples of research to mitigate potential harms include the following:

- Tools for mitigating bias and negative privacy impacts.
- Better approaches to system validation, inspectable models, and techniques for auditing system performance.
- Methods for anticipating and assessing the extent of, as well as approaches for addressing potential large or even extreme risks. A current example is assessing the consequences of potential biases or flaws in large-scale machine learning models when they are used as the basis for deployed AI-based systems affecting labor markets, physical processes, or financial practices.
- Research that deepens our understanding of the essential properties of responsible computing.
- Research that helps identify the limits of purely computing technical approach to societally harmful characteristics of computing technologies.

Recommendation 2.2. The computing research community should pursue, and research sponsors should invest in, computing research that would benefit the social good.

Examples of such research include the following: computing research aimed at such goals as sustainability, vaccine allocation, haptic feedback for robotic and telesurgery, or disaster planning and community resilience. Although there may be broad agreement on certain goals (e.g., sustainability or equity), the determination of social good inherently involves values and trade-offs and ultimately entails political decisions. It is the role of government and civil society to decide here as in many other circumstances on the social or public goods that they deem to be important. Research sponsors should look to neutral outside advisory groups for guidance from government, civil society, and economic actors about priorities. For NSF, the National Science Board could be the source of high-level guidance.

Recommendation 2.3. Computing research sponsors should foster and provide support for new kinds of projects that would help facilitate responsible computing research, including the multidisciplinary projects called for in Recommendation 1.

Potential new kinds of projects include the following:

- Government–computing research collaboratives that bring computing researchers (including students and other early career researchers) in direct contact with government agencies and civic and community organizations so that researchers better understand the contexts in which research results will be used and, potentially, contribute to the organization's technology needs and uses.
- Large-scale computing and data resources (e.g., the proposed National AI Research Resource) that would allow academic researchers to investigate capabilities that otherwise would only be available to large corporations. Such work must, of course, adhere to responsible computing guidelines.
- Public–private partnerships that support academic access to industry data in support of responsible computing research objectives. These partnerships can make possible research that neither academia nor industry could perform on their own. Such activities should include appropriate safeguards so that funding cannot be terminated, or publication restricted without adequate cause.

Recommendation 2.4. Research sponsors and universities should explore new partnerships with companies, nonprofits, and philanthropies to provide financial or in-kind support for responsible computing research with due attention to academic freedom.

One natural focus for such partnerships would be in addressing the sorts of challenges described in Recommendation 2.1. Possible models for such partnerships include the following:

- Research programs partially supported by one or more companies and managed by a federal funding agency, with such key elements as proposal selection and assessment performed by the federal agency. Another possibility is joint funding with a nonprofit organization, such as NSF's 2018 Early-Concept Grants for Exploratory Research on societal challenges arising from AI technology that was jointly supported by NSF and the Partnership on AI.[2] NSF has a significant track record with such partnerships.
- Arrangements by industry to provide researchers with access to such artifacts as open data sets.
- Industry–university collaborations that support industry sharing of ethical and societal impact challenges they have encountered and the processes and strategies they have used successfully to address them.
- Industry support for university-based research centers that emphasize responsible computing research and methods.

Recommendation 2.5. To enable carrying out Recommendation 1, research sponsors and research proposers should ensure that, where appropriate, awards provide sufficient resources to include researchers from fields outside computer science and to engage stakeholder groups and outside experts with societal, ethical, or domain expertise.

Recommendation 2.6. To enable carrying out Recommendation 1 and Recommendations 2.1 to 2.3, academic tenure and promotion committees and industry performance reviews should recognize the importance and value of scholarship, both disciplinary and multidisciplinary, that investigates the ethical and societal impacts of computing research.

---

[2] See National Science Foundation, 2018, "Dear Colleague Letter: Early-Concept Grants for Exploratory Research on Artificial Intelligence (AI) and Society—Supported Jointly with the Partnership on AI," NSF 19-018, https://www.nsf.gov/pubs/2019/nsf19018/nsf19018.jsp.

Responsible computing research requires multidisciplinary research, and the participation of researchers at all career stages in all relevant disciplines. The needed multidisciplinary research starts from an engagement with responsible computing issues and then identifies the concepts and reasoning from a relevant field that can be used to approach resolving them (e.g., "engaged ethics" as described in Section 2.1.1). If this research is to attract leading scholars and scientists, it must count in performance reviews. Just as tenure and promotion committees had to accommodate conference publications,[3] here too they need to adjust to accommodate the challenges of assessing the contributions of research that includes these highly multidisciplinary theorems.

4.3 SUPPORT THE DEVELOPMENT OF THE EXPERTISE NEEDED TO INTEGRATE SOCIAL AND BEHAVIORAL SCIENCE AND ETHICAL THINKING INTO COMPUTING RESEARCH

Recommendation 3. Universities, scientific and professional societies, and research and education sponsors should support the development of the expertise needed to integrate social and behavioral science and ethical thinking into computing research.

Responsible computing research requires that all participants in computing research possess a broader scope of expertise than is typical of most undergraduate majors or graduate programs. Responsible computing research will not significantly advance unless educational and training programs adapt and change.

Recommendation 3.1. Universities should enhance (1) teaching and learning in computer science and engineering, information science, and other computing-related fields to ensure that the next generation is better equipped to understand and address ethical issues and potential societal impacts of computing and (2) humanities and social and behavioral science education to ensure that students in those fields are equipped to participate in informed discussions of potential impacts of

---

[3] D. Patterson, L. Snyder, and J. Ullman, 1999, "Evaluating Computer Scientists and Engineers for Promotion and Tenure," *Computing Research News (September)*, http://archive2.cra.org/uploads/documents/resources/bpmemos/tenure_review.pdf.

computing research and technologies. Research sponsors should support such activities.

It is critical that these enhancements provide students with appropriate skills and experiences, rather than simply requiring students to take an introductory course in the other discipline(s). The goal here should be to teach students to understand ideas, theories, and results from the other discipline(s) and use them productively, rather than conduct or produce work in the other discipline(s).

The computing ethics projects described in the subsection "Integrating Ethical and Societal Issues into Training" in Chapter 3 are being carried out in diverse types of higher education settings, including small colleges, public and private universities, several of which serve populations underrepresented in computing fields. Their varying innovative approaches to integrating the teaching of ethics and responsible computing are leading indicators that academic institutions of all types can shape programs that address these needs. This variation also illustrates how each institution must, drawing on the various emerging models, make its own decisions about which approaches best fit their particular context.

Specific possible actions by universities include the following:

- More tightly integrate relevant ethics and social and behavioral knowledge into existing undergraduate and graduate computing courses and into curricular degree requirements. Sources of work that could be adopted or contributed to include the Online Ethics Center for Engineering and Science,[4] which has many resources for teachers that sometimes address sociotechnical systems and the Responsible Computer Science Challenge,[5] which has yielded a range of teaching materials, including curricular modules and other resources for integrating ethics and social sciences into the computer science curriculum,[6,7] the Computing Ethics Narratives project,[8] and the Massachusetts Institute of Technology's Case Studies in Social and Ethical Responsibilities of Computing.[9] Develop and share additional corpora of case studies, including examples encountered in research projects.

---

[4] See Online Ethics Center for Engineering and Science, "Homepage," https://onlineethics.org.
[5] A program sponsored by Omidyar Network, Mozilla, Schmidt Futures, and Craig Newmark Philanthropies.
[6] Mozilla, "Teaching Responsible Computing Playbook," https://foundation.mozilla.org/en/what-we-fund/awards/teaching-responsible-computing-playbook/teaching-materials.
[7] See Computing Ethics Narratives, "Homepage," https://www.computingnarratives.com.
[8] Ibid.
[9] D. Kaiser and J. Shah, 2021, *Case Studies in Social and Ethical Responsibilities of Computing,* Massachusetts Institute of Technology: MIT OpenCourseWare, https://ocw.mit.edu/resources/res-tll-007-case-studies-in-social-and-ethical-responsibilities-of-computing-fall-2021.

- Integrate a deeper knowledge of the powers and limitations of computing into undergraduate and graduate curricula in the social and behavioral sciences and humanities, thus enabling students to understand better ways in which their field can contribute to public understanding.
- Ensure that in particular those engaged in graduate training in one of the non-computing fields relevant to progressing responsible computing have the necessary technical expertise to subsequently collaborate and engage with technical computing disciplines.
- Provide opportunities for students to engage with computing research and its application in real-world social contexts; for example, through work with federal, state, and local governments.
- Design undergraduate majors, undergraduate- and master's-level certificates, or degrees in computing ethics for computer scientists and in computing for social and behavioral scientists and humanists. Existing examples include the minor in Societal and Human Impacts of Future Technologies at Carnegie Mellon University and the Master's in AI Ethics program at Cambridge University.
- Develop incentives—such as reduced course loads and support for course development—for faculty in both computer science departments and relevant social and behavioral sciences and humanities departments to collaborate in integrating the teaching of ethics and consideration of societal impact into the computer science curriculum.

Specific possible actions by sponsors of computing research and education include the following:

- Support the development and sharing of curricular materials by multidisciplinary teams and
- Explore ways to enable and encourage principal investigators (PIs) to expand the scope of educational programs to provide students with opportunities to participate in multidisciplinary research, whether through programs such as NSF's Research Experiences for Undergraduates or PI-directed educational activities. This may require new forms of such programs that can accommodate multiple kinds of faculty expertise in a grant.

Recommendation 3.2. Organizers of computing research conferences, workshops, and meetings of principal investigators should convene sessions or events at their meetings to share best practices and otherwise

promote responsible computing research, both disciplinary and multidisciplinary. Research sponsors should support such activities.

Specific actions in support of this recommendation include the following:

- Research sponsors should support academic, scientific, and professional organizations in hosting meetings of PIs and other computing researchers that expose them to effective and best practices for enabling deeply multidisciplinary research.
- Research sponsors should support substantive opportunities (e.g., summer intensive workshops) that bring together early career humanists, social and behavioral scientists, and computing researchers with established mentors in their respective disciplines focused on developing cohorts with deeper connections and literacies across the disciplines that need to cooperate on sociotechnical problems.

Recommendation 3.3. Conference organizers, journals, and research sponsors should provide computing researchers with guidelines and training opportunities on the appropriate ways to review ethical and societal issues in papers and proposals.

This guidance is particularly critical as researchers are increasingly required (e.g., by conferences and funding programs—see Recommendations 5.1, 5.3, and 6) to engage with ethical and societal issues, but typically lack the experience or training to appropriately meet these requirements. This recommendation focuses on the groups who are best positioned to provide much-needed training and education to researchers.

It is well established that peer review for multidisciplinary proposals faces particular challenges. Organizations that rely on peer review to assess academic quality should try to recruit scholars with experience of multidisciplinary work to review such proposals, rather than relying on experts in only one of the constituent disciplines.

Recommendation 3.4. Universities and computing research sponsors should, through their education and research activities, develop programs that help create the knowledge, expertise, and talent pool that public- and private-sector organizations will need to make knowledgeable decisions about their acquisition of computing technologies.

Decisions about acquisition and deployment are often made by individuals in government and the private sector who lack the training to assess the quality of a proposed acquisition with respect to its suitability and its potential impacts on all affected parties. This recommendation is meant to develop students with the expertise to realize Recommendation 8, below. Without a recommendation such as this one, computing research might become significantly more ethical and responsible even as computing use continues on its current problematic trajectory.

> **Recommendation 3.5. Research sponsors should support the development of techniques and capacity for evaluating the effectiveness of different approaches to enabling researchers to address ethical and societal implications of computing research.**

The computing research community has not yet been able to establish best practices for responsible computing research. Many ideas and proposals have been made, but there has been little practical implementation or empirical validation of these ideas. Moreover, research on methodological issues has historically not rewarded systematic empirical tests or validations. Hence, it is up to research sponsors to step into this gap to ensure that the research community is learning in a rigorous, informed manner how to do things better.

A specific opportunity for research sponsors is to support empirical tests of educational and workflow interventions to produce researchers and research teams who conduct more responsible computing research—for example, research to systematically examine the effectiveness and other implications of different models for ethics review of research projects and scholarly publications.

Research sponsors, institutions participating in research, and in some cases external civil society organizations can all contribute by developing the capacity to assess, and potentially audit, the extent to which these ethical and societal concerns are being taken seriously in research over time.

4.4 ENSURE THAT RESEARCHERS HAVE ACCESS TO THE KNOWLEDGE AND EXPERTISE NEEDED TO ASSESS THE ETHICAL AND SOCIETAL IMPLICATIONS OF THEIR WORK

> **Recommendation 4. Computing research organizations—working with scientific and professional societies and research sponsors—should ensure that their computing faculty, students, and research staff have**

access to scholars with the expertise to advise them in examining potential ethical and societal implications of proposed and ongoing research activities, including ways to engage relevant groups of stakeholders. Computing researchers should seek out such advice.

Many of the recommendations in this report (notably Recommendations 1, 2, and 5) assume that computing researchers can access the expertise in ethics and social and behavioral sciences they need to design and carry out responsible research projects. This recommendation addresses several ways that such expertise can be made available in a variety of institutional settings.

Computing researchers should seek out such advice within their own institutions and from scientific and professional organizations, suitable expertise on the ethical and societal implications of their research, and if such expertise is not immediately available, work with their institutions and networks to develop access to it.

Recommendation 4.1. Computing research organizations should identify in-house experts who can be consulted by computing faculty, students, and research on how to address ethical considerations and societal impacts of their research. As scholars from most of the disciplines that computing researchers will need to draw on would need funding to collaborate, universities should work with the various stakeholders to help identify resources that can enable this collaboration.

Importantly, this recommendation does not call for creating another mechanism for centralized institutional review of research project designs but rather to create the capacity for ongoing consultation by researchers as projects evolve. Such an approach mirrors practices emerging in industry to bring early consultation into earlier stages of research and design. There is a useful analogy to statistical consulting centers that many universities have. An important goal for such centers is to provide opportunities for researchers to consult experts in statistics in the early phases of their research project ideation, when larger changes are most readily made. This analogy also suggests potential funding models, such as a mix of institutional support for shorter-term or smaller-scale engagements and support from research project budgets for larger engagements.

Recommendation 4.2. Scientific and professional societies of computing researchers should help their members identify experts whom they can consult when developing proposals and carrying out research

projects. **Such connections can be particularly valuable for researchers who do not have extensive access to in-house expertise at their own institutions.**

Smaller institutions may not have the capacity to provide in-house expertise in responsible computing research. Early career stage computing professionals are least likely to know experts in relevant non-computing fields, and they might be more comfortable engaging with experts outside their institution who are familiar with their own field or subfield of computing research.

Recommendation 4.3. Research sponsors and scientific and professional societies should support the development and sharing of tutorials, best practice descriptions, class materials, and the like to provide concrete examples of solutions that have worked previously for researchers.

The best way to speed progress is to share best practices and other knowledge as widely as possible throughout the research community.

4.5 INTEGRATE ETHICAL AND SOCIETAL CONSIDERATIONS INTO COMPUTING RESEARCH SPONSORSHIP

Recommendation 5. Sponsors of computing research should require that ethical and societal considerations be interwoven into research proposals, evaluated in proposal review, and included in project reports.

This recommendation recognizes the sociotechnical nature of computing research. It is designed to ensure that engagement with ethical and social responsibility becomes a routine component of computing research throughout its lifecycle, starting with research ideation and design, continuing with the challenges encountered and changes to project plans that occur as research projects evolve, and ending with reporting of lessons learned from the research.

Note that this recommendation and the subrecommendations under it call for something distinct from and complementary to existing institutions, rules, and practices designed to protect human subjects (e.g., institutional review boards and informed consent rules) or to ensure the ethical conduct of research (e.g., the National Science Foundation's requirements for responsible and ethical conduct of research[10]).

Recommendation 5.1. Computing research proposals should describe in an integrated fashion the ethical and societal considerations associated with the proposed work and ways the work will address those considerations. Research sponsors should avoid requiring a freestanding top-level section focused on ethics because such a separation from the main contents of the proposal would treat ethics as an add-on rather than integral to the proposed work.

Specifically:

- The content of computing research proposals should describe the ethical and societal considerations associated with the proposed work and how they will be addressed in the proposed research (e.g., participation of researchers from other disciplines or involvement of domain experts and stakeholders).
- Ethical and societal considerations should be integrated into the main body of the proposal (e.g., addressed explicitly in sections describing the proposed work, its intellectual merit, and its broader impacts) and not segregated into an "add-on" section.
- The proposed list of researchers who will participate in carrying out the research should include the full range of requisite disciplinary expertise, and the proposed project plan and budget should reflect their active participation.

Implementers of this recommendation can draw on the experiences of recent experiments with oversight of responsible computing research at the proposal stage. Two recent examples are the Stanford Institute for Human-Centered Artificial Intelligence's Ethics and Society Review board[11] and the Microsoft Research Ethics Review Program.[12] There are other interesting examples of ethics oversight in the corporate sector that are in service of product development rather than research and are not addressed here.

---

[10] National Science Foundation, "Responsible and Ethical Conduct of Research," https://www.nsf.gov/od/recr.jsp.

[11] M.S. Bernstein, M. Levi, D. Magnus, B. ARajala, D. Satz, and C. Waeiss, 2021, "ESR: Ethics and Society Review of Artificial Intelligence Research," arXiv, https://arxiv.org/abs/2106.11521.

[12] Microsoft, "Microsoft Research Ethics Review Program & IRB," https://aka.ms/msrethicsreviewprogram.

Recommendation 5.2. Computing research sponsors should develop criteria for evaluating ethical and societal considerations and ensure that project review panels have the requisite expertise to conduct such evaluations.

Individual computer scientists may not currently possess the expertise to consistently evaluate research project designs for the adequacy of their presentation of ethical and societal impacts. Thus, a first step for the computing research community is to establish guidelines for evaluation of proposals to assess whether they appropriately address ethics concerns specific to the field of computer science. These criteria will also serve as a signal of what responsible computing research looks like to researchers developing proposals. Chapters 2 and 3 of this report are intended to serve as a useful source in developing such criteria.

Specifically:

- Funding agencies should develop and publicize transparent criteria for evaluation of ethical and societal impact considerations (e.g., in the descriptions of the research, its intellectual merit and broader impacts).
- To have the requisite expertise on review panels, the panels should (1) include researchers with relevant expertise in assessing how well a proposal takes into account ethical and societal impacts; (2) be diverse and inclusive (e.g., to increase the likelihood that impacts on all relevant communities are considered); and (3) include reviewers with significant experience participating in multidisciplinary projects.

Recommendation 5.3. Computing research sponsors should require that project reports address ethical and societal issues that arose.

Research is dynamic and emergent, and researchers often encounter unexpected roadblocks and results. In response, researchers adapt and adjust the course of their projects. As they do so, they need also to reassess and modify their plans for handling potential ethics and societal impact effects of the work. Researchers should periodically update sponsors about the shifts in this aspect of their projects. To avoid additional reporting requirements, the most appropriate place to address adaptations and updates to these changes as well as to the technical research plans is in existing reporting mechanisms.

A concern, particularly for young researchers or researchers who might feel vulnerable, is that such reports to program officers about challenges or changes in research design could have career-affecting negative consequences—for example, lead to exclusion from future funding opportunities. Thus, it will be important that any response to such reports be to provide non-judgmental guidance to researchers as they responsibly attempt to adjust their research plans.

Specifically:

- Research sponsors should require that interim project reports provide updates on societal or ethical issues encountered and describe the ways they were addressed.
- Research sponsors should require that final project reports discuss lessons learned about ethical and societal impact. In particular, they should include a summary of any unanticipated ethical or social consequences of the research and provide guidance regarding ethical considerations to future researchers and developers who might subsequently extend or use the results of the research.
- Research sponsors should develop ways to share high-level, salient lessons learned in an anonymized, aggregated manner.

Recommendation 5.4. Research sponsors should evaluate the impacts of their implementation of these recommendations after approximately 5 years (i.e., at least one grant cycle after new requirements are developed and issued) and periodically thereafter.

Recommendations 5.1 to 5.3 are, in the view of the committee, the plausible, practical steps that research sponsors could take to ensure that the computing research they support addresses ethical and societal impact challenges. The intent is that these recommendations will lead to deep and effective engagement with ethical and societal impacts by the research community yet not be overly burdensome to researchers or their sponsors. Because the proposed interventions are new, their effectiveness should be assessed, and the interventions adapted as needed to meet the intention of the recommendation.

4.6 INTEGRATE ETHICAL AND SOCIETAL CONSIDERATIONS INTO PUBLICATION

Recommendation 6. Scientific and professional societies and other publishers of computing research should take steps to ensure that ethical and societal considerations are appropriately addressed in publications. The computing research community should likewise take steps to ensure that these considerations are appropriately addressed in the public release of artifacts.

This recommendation is intended to be applied to all the ways in which computing research is published or otherwise released. In many areas of computing research, conference proceedings are at least as important as journals in publishing research results. Moreover, computing research also produces artifacts—code, models, and data—that play an important and complementary role to conferences and journals in disseminating research results. For example, software is often released with an open-source license and deposited in a repository such as GitHub. And like many other disciplines, computing researchers frequently deposit papers in non-reviewed repositories such as arXiv. The recommendations in this report call on computing researchers and organizations where computing research is conducted to address responsibility not only in conferences and journals but also when releasing artifacts and papers in repositories such as GitHub and arXiv.

Many subfields of computing research have robust traditions of releasing code and data as part of scientific publications, to ensure reproducibility and, in many cases, to foster the wider use of these artifacts. Some of these research artifacts develop large user bases over time, numbering in the tens of thousands. Standard conference and journal review processes often omit in-depth examination of accompanying artifacts; as such artifacts are critical to the field (and potentially to society), they should be treated as first-class objects in scientific publications.

Recommendation 6.1. Conferences and journals should include in their evaluation criteria and metrics an assessment of how well a paper addresses ethical issues and societal impacts associated with the research, approaches taken by the researchers to mitigate these issues, and potential approaches that future researchers or developers using these results should take to mitigate potential negative impacts.

Specifically:

- Publication venues should establish a clear set of responsible computing research guidelines for authors and reviewers. These guidelines should be created by the venue's governing scientific or professional association[13] or created by the venue after consultation with a diverse group of scientists. (These guidelines would be in addition to whatever they may require with respect to protecting human research subjects.)
- The guidelines adopted by the publication venue should be posted publicly and linked from the call for papers.

Recommendation 6.2. Conferences and journals should encourage researchers to report unanticipated ethical or social consequences of the research as well as guidance regarding ethical considerations to other researchers and developers who might use the research in the future.

Specifically:

- Publication venues should provide authors with space within the main paper page limits to discuss the ethical considerations of their research.
- Publication venues should provide authors and reviewers questions to consider, "model" ethical considerations discussions, or similar sets of examples and guidelines.[14]
- Publication venues should adopt procedures for paper withdrawal in cases where it has been reliably established that there is inadequate attention to ethical or societal consequences of the reported research.
- Publication venues should consider adopting procedures for post-publication discussion of papers' ethical or societal impacts. A model is provided by journals, for example, in the behavioral sciences, that solicit and publish comments on papers.

---

[13] For example: Linguistic Society of America, "Ethics," https://www.linguisticsociety.org/resource/ethics; ACM Ethics, "2018 ACM Code," https://ethics.acm.org/; IEEE, "IEEE Code of Ethics," https://www.ieee.org/about/corporate/governance/p7-8.html; Association of Computational Linguistics, "ARR ACL Rolling Review," https://aclrollingreview.org/responsibleNLPresearch/; Neural Information Processing Systems, "A Retrospective on the NeurIPS 2021 Ethics Review Process," https://blog.neurips.cc/2021/12/03/a-retrospective-on-the-neurips-2021-ethics-review-process.

[14] See P. Nanyakkara, J. Hullman, and N. Diakopoulos, 2021, "Unpacking the Expressed Consequences of AI Research in Broader Impact Statements." Poster Paper Presentation, Pp. 795-806 in AIES '21, May 19-21, 2021, Virtual Event, https://drive.google.com/file/d/1-sj-jP4VNYHClajOLo0pSMk6HxYimh1E/view?usp=sharing. This paper analyzes broader impact statements in NeurIPS papers and suggests ways to better frame them.

Recommendation 6.3. Review committees should use these criteria and metrics and possess the multidisciplinary expertise needed to do so.

Specifically:

- Publication venues should consider whether to assign ethics review as a standard reviewer responsibility or to assign ethics review to a separate set of reviewers.
- If ethics review is a standard reviewer responsibility, reviewers need to be educated about the responsible computing research guidelines adopted by the venue.
- If ethics review is assigned to a separate set of reviewers, care should be taken to ensure the pool of ethics reviewers is diverse (e.g., geographically diverse) and informed about the guidelines adopted by the venue.

Recommendation 6.4. Scientific and professional societies should identify experts who are willing to be consulted by paper authors, paper reviewers, and program committees.

Analogous to Recommendations 4.1 and 4.2, which are concerned with performing research, preparing and reviewing publications may also require the involvement of people whose primary expertise is not in the field of the publication venue. For example, an AI venue may need to draw on the expertise of a medical ethicist for certain submissions.

Recommendation 6.5. Scientific and professional societies should, with participation and support from academia and industry research institutions, establish criteria for whether and how to release artifacts (hardware, code, models, or data sets) that may have harmful effects. Released artifacts should be accompanied by information about their intended uses, limitations, and potential harmful effects.

There are recently invigorated norms around making artifacts public, especially when they result from publicly funded research. One reason is reproducibility, which is a core scientific value, and in many cases, artifacts are critical to reproducibility. Another is to make the results of research available widely so as to fuel further innovation. However, full transparency must be balanced against other societal values, including but not limited to privacy (data that contains personally identifiable information, where the research participants did not consent to data release), intellectual property (data, code, or models), and security (data, code, or models that have the potential to weaken national security or that expose significant vulnerabilities in widely used products).

With these points in mind, publication venues should encourage authors and reviewers to think critically and carefully about whether and how to release research artifacts. There are multiple options for releasing artifacts including the following:

- Unrestricted access and use,
- Restrictive licenses,
- Limits on access and monitoring of use, and
- Release of a weakened version of the artifact that mitigates its harmful effects while still allowing others to reproduce or build on the research results.

Code may be buggy; data may be biased or incomplete; models may be incorrect. In certain circumstances, authors may decide not to release artifacts described in their papers. Reviewers should not reject papers for the sole reason that they are not accompanied by research artifacts. Publication venues should consider deeper notions of reproducibility than merely the release of code and other artifacts.[15]

Publication venues that host or link to research artifacts other than papers (e.g., code, data) should:

- Consider creating "model" or template code readmes, data sheets, model cards, and so on;
- Encourage reviewers to examine and comment on artifacts that are submitted with papers, or consider separate research artifact review; and
- Define and adopt procedures for artifact withdrawal or update, should authors or others identify ethical failures or gaps.

Recommendation 6.6. Computing research conferences should adopt policies and principles governing when they accept sponsorship and make these policies publicly available.

---

[15] Venues that want to encourage reproducibility can host reproducibility hackathons (F. Balsiger, A. Jungo, N.A. R J, J. Chen, I. Ezhov, S. Liu, J. Ma, et al., "The MICCAI Hackathon on Reproducibility, Diversity, and Selection of Papers at the MICCAI Conference," arXiv, https://arxiv.org/abs/2103.05437); publish interesting negative results, "Journal of Interesting Negative Results in Natural Language Processing and Machine Learning," http://jinr.site.uottawa.ca); and provide reproducibility checklists of authors and reviewers (Association for the Advancement of Artificial Intelligence, "Reproducibility Checklist," https://aaai.org/Conferences/AAAI-21/reproducibility-checklist).

Many computing research conferences accept financial support from outside sponsors. Many of these conferences do not have a policy stating the conditions governing such support. Without such a policy, conference organizers handle controversies that arise on an ad hoc basis. By establishing a policy and making it public, conference organizers will save time and energy and the process will be (and be seen as) fairer. One example of such a policy is the "Sponsorship Policy of the ACM Conference on Fairness, Accountability, and Transparency."[16]

4.7 ADHERE TO BEST PRACTICES FOR SYSTEMS DESIGN, DEPLOYMENT, OVERSIGHT, AND MONITORING

Recommendation 7. Computing researchers who are involved in the development or deployment of systems should adhere to established best practices in the computing community for system design, oversight, and monitoring.

Computer science and information science and engineering scholarship along with best practices developed in industry provide a wealth of information about such practices.

Recommendation 7.1. Researchers should follow all well-established best practices for system design.

There are many such best practices for designers and developers including the following:

- For algorithms that optimize, consider the need to optimize for human expertise or the complementarity of human and machine expertise rather than for machine expertise alone;
- Make systems accessible regardless of such differences in abilities as cognitive (including literacy), visual, motor, or hearing ones and regardless of cultural background (e.g., first language, dialect, or accent);
- Integrate technology with organizational practices;
- Make system behaviors and results transparent to the full range of their users and provide automated tooling (e.g., visualizations and performance analysis

---

[16] See Association for Computing Machinery, "ACM FAccT Conference Sponsorship Policy," https://facctconference.org/sponsorship.html.

tools) so users can understand what a system is doing and how it is interacting with its environment;

- During the system design stage, conceptualize the success of the system not just in terms of quantitative metrics like "performance" or "accuracy," but also with respect to broader issues involving ethical concerns and broader societal impacts, even if those factors admit only to qualitative, not quantitative assessment;
- Involve appropriately diverse expertise and stakeholders, ensuring that people and groups affected by computing systems are involved in their design, taking into account the global reach of computing technology;
- Ensure that the security and privacy of data is considered at design time, and that these considerations touch every aspect of the system; and
- Identify potential unanticipated uses and mitigate the harms they could cause.

Recommendation 7.2. Researchers producing artifacts (hardware, code, models, and data sets) should be sufficiently transparent as the artifact evolves during their research about the capabilities, maturity, limitations, and potential ethical and societal impacts of the artifacts so that researchers building systems and vendors building products incorporating the artifact and users of those products can adequately assess them. This transparency should be maintained as the artifact evolves.

The intention of this recommendation is that researchers provide enough information so that other researchers and vendors incorporating those artifacts can validate systems prior to and after deployment. Doing so is essential for vendors themselves to be transparent about their systems and allow for vendor-independent evaluation by regulators (where applicable), purchasers, and users. In doing so, it is important for all parties to be clear whether an artifact has been released at an early stage (i.e., not fully tested) in order to get feedback. One recent effort at such transparency can be found in a paper discussing risks of foundation models.[17] Before a full public release, researchers should seriously consider engaging with a diverse group of potential users.

---

[17] R. Bommasani, D.A. Hudson, E. Adeli, R. Altman, S. Arora, S. von Arx, M.S. Bernstein, et al., 2021, "On the Opportunities and Risks of Foundation Models," https://doi.org/10.48550/arxiv.2108.07258.

Recommendation 7.3. Researchers should recognize the potential lifetimes of computing systems that may be built based on their work. They should document their design assumptions and, if possible, build in safeguards for triggering reassessment of these design assumptions.

In practice, computing systems are evolving constructions that require adaptation in response to changing facts. Such adaptations should trigger a reconsideration of potential ethical and societal impacts.

4.8 SUPPORT ENGAGEMENT WITH THE PUBLIC AND THE PUBLIC INTEREST

Recommendation 8. Research sponsors, research institutions, and scientific and professional societies should encourage computing researchers to engage with the public and with the public interest and support them in doing so.

Individuals and societies as a whole are often affected by new technologies and would benefit from opportunities to better understand what's going on "under the hood." Note that public engagement (Recommendation 8.1) and transparency (Recommendation 8.2) are necessary but not sufficient: Recommendation 8.3 helps provide information that lawmakers, regulators, and other decision-makers need to make informed governance decisions.

Recommendation 8.1. Researchers should consider proposing, and research sponsors should consider supporting, public engagement activities at relevant stages of research projects to inform the public about emerging computing technologies.

Specific opportunities include the following:

- Include multidisciplinary expertise on research teams to help identify relevant publics and effective outreach strategies,
- Leverage existing institutional communications capabilities, and
- Support the involvement of members of the public where needed.

Recommendation 8.2. Computing researchers should develop and promulgate knowledge that supports better decision-making by acquirers of computing technologies, particularly governments. Research sponsors should consider supporting such work as well as research on effective ways of informing decision-makers and the public.

Past mistakes could potentially have been avoided if such information had been available to technology developers or the governments and other users that adopted the technology. For instance, government agencies adopted face recognition technology or systems for parole decision-making only to discover serious bias issues after deployment. Potential opportunities include:

- Develop methodologies for creating effective "buyers guides" and "users guides" (a data and computing system equivalent of "good features of" nutrition and drug labels) for computing technologies;
- Advise on appropriate approaches for evaluating computing technologies applied in socially impactful contexts;
- Advise on governance gaps and challenges as well as potential approaches; and
- Better document parameters and conditions of applicability and appropriate use of research project results (algorithms, code, and systems) and make them available to decision-makers and users.

Recommendation 8.3. Universities, research funding agencies, and scientific and professional societies should provide opportunities for computing researchers (along with their collaborators in other disciplines and application domains) to advise the public about the limitations as well as the strengths of emerging computing technologies and provide settings in which researchers can learn how to serve effectively in advisory capacities.

Potential opportunities include programs such as the National Academies' Jefferson Science Fellowship and the American Association for the Advancement of Science's Science and Technology Fellowships, which bring scientists and engineers to work in federal agencies, and science communication programs of such organizations as the Kavli Foundation and the Alda Center for Communicating Science at Stony Brook University.

Recommendation 8.4. Computing researchers (along with their collaborators in other disciplines and application domains) should be encouraged by universities, research funding agencies, and scientific and professional societies to bring their knowledge of potential effects and consequences to governments and civil society organizations early in a technology's development and as a technology is considered for or used in different contexts so that potential negative consequences of that technology can be understood and mitigated adequately.

This recommendation aims to provide public officials with greater insights as to the need for regulation of existing or new technologies. As new technologies are developed, some research sponsors may want to fund research on whether and where new regulations may be needed. Importantly, in addition to permitting "good regulation" it helps avoid "bad regulation," including potential stifling of innovation. Disclosure allows for feedback from in-house and external technologists, avoiding bad regulation (and perhaps stifling of innovation) and permitting good regulation.

Appendixes

A
Committee Member Biographical Information

BARBARA J. GROSZ, *Chair,* is the Higgins Research Professor of Natural Sciences in the Paulson School of Engineering and Applied Sciences at Harvard University. Dr. Grosz has made groundbreaking contributions to the field of artificial intelligence through her pioneering research in natural language processing and in theories of multi-agent collaboration and their application to human–computer interaction. Her recent research has explored ways to use models developed in this research to improve health care coordination and science education. Dr. Grosz is also known for her role in the establishment and leadership of multidisciplinary institutions and is widely respected for her many contributions to the advancement of women in science. She co-founded Harvard's Embedded Ethics program, which integrates teaching of ethical reasoning into core computer science courses. As the founding dean of science and then the dean of Harvard's Radcliffe Institute for Advanced Study, she designed and launched the Institute's science program and subsequently its Academic Ventures program. She was the founding chair of the Standing Committee for Stanford's One Hundred Year Study on Artificial Intelligence and has served on the boards of several scientific, scholarly, and academic institutions, including the National Academies of Sciences, Engineering, and Medicine's Computer Science and Telecommunications Board from 1994–1998. A member of the American Philosophical Society, Dr. Grosz is a fellow of the American Academy of Arts & Sciences, the Association for the Advancement of Artificial Intelligence (AAAI), the Association for Computing Machinery (ACM), and the American Association for the Advancement of Science, and a corresponding fellow of the Royal Society of Edinburgh. She received the 2009 ACM/AAAI Allen Newell Award, the 2015 International Joint Conference on Artificial Intelligence Award for Research Excellence, and the 2017 Association for Computational

Linguistics Lifetime Achievement Award. In 2017, Dr. Grosz received an Everett Mendelsohn Excellence in Mentoring Award from Harvard's Graduate Student Council. Dr. Grosz received an A.B. in mathematics from Cornell University and a Ph.D. in computer science from the University of California, Berkeley.

MARK ACKERMAN is the George Herbert Mead Collegiate Professor of Human–Computer Interaction and a professor in the School of Information in the Department of Electrical Engineering and Computer Science at the University of Michigan, Ann Arbor. His major research area is human–computer interaction (HCI), primarily computer-supported cooperative work (CSCW). He has published widely in HCI and CSCW, investigating collaborative information access in online knowledge communities, medical settings, expertise sharing, and most recently, pervasive environments. Dr. Ackerman is a member of the CHI Academy (HCI Fellow) and an Association for Computing Machinery Fellow. Previously, Dr. Ackerman was a faculty member at the University of California, Irvine, and a research scientist at the Massachusetts Institute of Technology's (MIT's) Laboratory for Computer Science (now CSAIL). Before becoming an academic, Dr. Ackerman led the development of the first home banking system, had three Billboard Top-10 games for the Atari 2600, and worked on the X Window System's first user-interface widget set. Dr. Ackerman has degrees from the University of Chicago, The Ohio State University, and MIT. Dr. Ackerman received his M.S. in computer science from The Ohio State University and his Ph.D. in information technologies from MIT.

STEVEN M. BELLOVIN is the Percy K. and Vidal L.W. Hudson Professor of Computer Science at Columbia University, a member of the Cybersecurity and Privacy Center of the university's Data Science Institute, and an affiliate faculty member at Columbia Law School. He does research on security and privacy and on related public policy issues. He received a B.A. from Columbia University, and an M.S. and a Ph.D. in computer science from the University of North Carolina at Chapel Hill. Dr. Bellovin has served as the chief technologist of the Federal Trade Commission and as the Technology Scholar at the Privacy and Civil Liberties Oversight Board. He is a member of the National Academy of Engineering and has served on the National Academies of Sciences, Engineering, and Medicine's Computer Science and Telecommunications Board. In the past, he has been a member of the Department of Homeland Security's Science and Technology Advisory Committee and the Technical Guidelines Development Committee of the Election Assistance Commission.

MARIANO-FLORENTINO (TINO) CUÉLLAR is the 10th president of the Carnegie Endowment for International Peace. A former justice of the Supreme Court of California, he served two U.S. presidents at the White House and in federal agencies and was a faculty member at Stanford University for two decades. Before serving on California's highest court, he was the Stanley Morrison Professor of Law, a professor (by courtesy) of political science, and the director of the Freeman Spogli Institute for International Studies at Stanford. A fellow of the American Academy of Arts & Sciences, Dr. Cuéllar is the author of *Governing Security: The Hidden Origins of American Security Agencies* (2013) and has published widely on transnational regulatory and security problems, American institutions, public law, and technology's impact on law and government. Dr. Cuéllar co-authored the first ever report on the use of artificial intelligence across federal agencies. He has served on the American Academy of Arts & Sciences Commission on Accelerating Climate Action. He chairs the board of the William & Flora Hewlett Foundation and is a member of the Harvard Corporation. Born in Matamoros, Mexico, he grew up primarily in communities along the U.S.-Mexico border. He graduated from Harvard College and Yale Law School and received a Ph.D. in political science from Stanford University.

DAVID DANKS is a professor of data science and philosophy and affiliate faculty in the Department of Computer Science and Engineering at the University of California, San Diego. Previously he was a professor of philosophy and psychology, the head of the Department of Philosophy, the chief ethicist of the Block Center for Technology and Society, and the co-director of the Center for Informed Democracy and Social Cybersecurity at Carnegie Mellon University. Dr. Danks is the recipient of a James S. McDonnell Foundation Scholar Award and an Andrew Carnegie Fellowship. He received an A.B. in philosophy from Princeton University and a Ph.D. in philosophy from the University of California, San Diego.

MEGAN FINN is an associate professor at the University of Washington Information School. She published the monograph *Documenting Aftermath: Information Infrastructures in the Wake of Disasters* about post-earthquake communication practices. Her newer projects examine ethical research practices in the field of computer security and investigate the implications of novel information policies on a transnational scale. She brings together perspectives and approaches from information studies; science and technology studies; and the history of media, information, and communication. In addition to her research and teaching, she is an advisor for the Science, Technology, and Society Studies Graduate Certificate program, a member of the iSchool's DataLab, and starting in 2019, a core faculty in Data Science Studies at the eScience Institute. Dr. Finn has an undergraduate degree in computer science from the University of Michigan, completed her

Ph.D. at the University of California, Berkeley, and spent time at Microsoft Research New England in Cambridge, Massachusetts, with the Social Media Collective as a postdoctoral researcher.

MARY L. GRAY is a senior principal researcher at Microsoft Research and a faculty associate at Harvard University's Berkman Klein Center for Internet and Society. She also maintains a faculty position in the School of Informatics, Computing, and Engineering with affiliations in anthropology and gender studies at Indiana University. Dr. Gray, an anthropologist and a media scholar by training, focuses on how everyday uses of technologies transform people's lives. Dr. Gray is the author, with computer scientist Siddharth Suri, of *Ghost Work: How to Stop Silicon Valley from Building a New Global Underclass*, published by Houghton Mifflin Harcourt in 2019. It was named a Financial Times' Critic's Pick and awarded the McGannon Center for Communication Research Book Prize in 2019. Her other books include *In Your Face: Stories from the Lives of Queer Youth*, *Queering the Countryside: New Directions in Rural Queer Studies*, a 2016 Choice Academic Title co-edited with Colin Johnson and Brian Gilley, and *Out in the Country: Youth, Media, and Queer Visibility in Rural America*, which looked at how young people in rural Southeast Appalachia use media to negotiate their sexual and gender identities, local belonging, and connections to broader, imagined queer communities. Dr. Gary received her Ph.D. in communications from the University of California, San Diego.

JOHN L. HENNESSY is the chair of Alphabet, Inc.; a professor of electrical engineering and computer science at Stanford University; and the director of Stanford's Knight-Hennessy Scholars Program. Dr. Hennessy served as the president of Stanford University from September 2000 until August 2016. Dr. Hennessy, a pioneer in computer architecture, joined Stanford's faculty in 1977 as an assistant professor of electrical engineering. In 1981, he drew together researchers to focus on a technology known as RISC (Reduced Instruction Set Computer), which revolutionized computing by increasing performance while reducing costs. Dr. Hennessy helped transfer this technology to industry co-founding MIPS Computer Systems in 1984. He is the co-author (with David Patterson) of two internationally used textbooks in computer architecture. His honors include the 2012 Medal of Honor of the Institute of Electrical and Electronics Engineers, the 2017 Association for Computing Machinery (ACM) Turing Award (jointly with David Patterson), the 2001 Eckert-Mauchly Award of ACM, the 2001 Seymour Cray Computer Engineering Award, and the 2004 NEC C&C Prize for lifetime achievement in computer science and engineering. He is an elected member of the American Academy of Arts & Sciences, the Royal Academy of Engineering, and the American Philosophical Society and has served on a number of National Academies of Sciences, Engineering, and Medicine boards and committees. Dr. Hennessy earned a Ph.D. in computer science from Stony Brook University.

AYANNA M. HOWARD is the dean of The Ohio State University College of Engineering and a professor in the college's Department of Electrical and Computer Engineering with a joint appointment in computer science and engineering. Previously she was the chair of the Georgia Institute of Technology School of Interactive Computing in the College of Computing, as well as founder and director of the Human-Automation Systems Lab (HumAnS). Dr. Howard is the founder and the president of the board of directors of Zyrobotics, a Georgia Tech spin-off company that develops mobile therapy and educational products for children with special needs. From 1993 to 2005, she worked at NASA's Jet Propulsion Laboratory, where she held multiple roles including senior robotics researcher and deputy manager in the Office of the Chief Scientist. Dr. Howard earned her bachelor's degree in computer engineering from Brown University, her master's degree and Ph.D. in electrical engineering from the University of Southern California, and her M.B.A. from Claremont Graduate University.

JON M. KLEINBERG is a professor in both computer science and information science at Cornell University. His research focuses on issues at the interface of networks and information, with an emphasis on the social and information networks that underpin the Web and other on-line media. His work has been supported by a National Science Foundation (NSF) Career Award, an Office of Naval Research Young Investigator Award, a MacArthur Foundation Fellowship, a Packard Foundation Fellowship, a Sloan Foundation Fellowship, and grants from Google, Yahoo!, and NSF. He is a member of the National Academy of Sciences, the National Academy of Engineering, and the American Academy of Arts & Sciences. Dr. Kleinberg received a B.S. in computer science from Cornell University in 1993 and a Ph.D., also in computer science, from the Massachusetts Institute of Technology in 1996.

SETH LAZAR is a professor in the School of Philosophy at the Australia National University, the lead investigator on the Australian Research Council grant "Ethics and Risk," the director of a Templeton World Charity Foundation project on "Moral Skill and Artificial Intelligence," and the project leader of the major interdisciplinary research project: Humanising Machine Intelligence. In 2019, he was awarded the Australian National University Vice Chancellor's award for excellence in research. His first book, *Sparing Civilians* (Oxford, 2015), aims to preserve the protection of civilians in war against political and philosophical threats that have arisen in recent years. A central focus of his early work on the ethics of war was the necessity of taking an approach more grounded in political philosophy than in moral philosophy—the same redirection is necessary for work on the morality, law, and politics of data and artificial intelligence. He has published papers in many top journals, including *Ethics* (2009, 2015, 2017), *Philosophy and Public Affairs* (2010, 2012, 2018), *Australasian Journal of Philosophy* (2015), *Nous* (2017), *Synthese* (2019), *Philosophical Quarterly* (2018), *Philosophical Studies* (2017), *Oxford Studies in Political Philosophy* (2017), and others. Dr. Lazar received his Ph.D. in political theory from Oxford University in 2009.

JAMES MANYIKA is a senior vice president at Google, Inc., and a senior partner emeritus and the chair emeritus of the McKinsey Global Institute. At MGI, Dr. Manyika has led research on technology, future of work, productivity, and economic growth. He was appointed the vice chair of the Global Development Council at the White House by President Obama and appointed by U.S. Commerce Secretaries to serve on the National Innovation Advisory Board and the Commerce Department's Digital Economy Board of Advisors. He is a visiting professor in technology and governance at Oxford University's Blavatnik School of Government. He serves on the boards of Council on Foreign Relations, MacArthur Foundation, Hewlett Foundation, the Broad Institute of the Massachusetts Institute of Technology (MIT) and Harvard, and research advisory boards at MIT, Harvard, Oxford, and Stanford, including as a member of the steering committee of the 100-year study on artificial intelligence (AI). He earned D.Phil., M.Sc., and M.A. degrees in AI and robotics, mathematics, and computer science from Oxford as a Rhodes Scholar. He is a fellow of the American Academy of Arts & Sciences, a fellow of the Royal Society of Arts, a Distinguished Fellow of Stanford's AI Institute, a Distinguished Research Fellow in Ethics and AI at Oxford, a fellow of DeepMind. He was a visiting scientist at NASA's Jet Propulsion Laboratory, and a faculty exchange fellow at MIT. At Oxford, he was a member of the Programming Research Group, the Robotics Research Lab, and elected a research fellow of Balliol College.

JAMES MICKENS is the Gordon McKay Professor of Computer Science at Harvard University's John A. Paulson School of Engineering and Applied Sciences. His research focuses on distributed systems, such as large-scale services, and ways to make them more secure. He joined the Distributed Systems group at Microsoft Research in 2009, and Harvard's School of Engineering and Applied Sciences in 2015, where he was awarded tenure in 2019. Dr. Mickens received a Ph.D. in computer science and engineering from the University of Michigan in 2008 and his B.S. in computer science from the Georgia Institute of Technology in 2001.

AMANDA STENT is the inaugural director of Colby College's Davis Institute for Artificial Intelligence. Previously, she was the natural language processing architect at Bloomberg L.P. and led the company's People and Language AI Team. Before that, she was a director of research and the principal research scientist at Yahoo Labs, a principal member of technical staff at AT&T Labs-Research, and an associate professor in the Computer Science Department at Stony Brook University. Her research interests center on natural language processing and its applications, in particular topics related to text analytics, discourse, dialog, and natural language generation. She is the co-editor of the book *Natural Language Generation in Interactive Systems* (Cambridge University Press), has authored more than 90 papers on natural language processing and is the co-inventor on more than 20 patents and patent applications. She is the president emeritus of the Association of Computational Linguistics (ACL)/International Speech Communication Association Special Interest Group on Discourse and Dialog, treasurer of the ACL Special Interest Group on Natural Language Generation, and one of the rotating editors of the journal *Dialogue and Discourse*. She holds a Ph.D. in computer science from the University of Rochester.

B

Presentations to the Committee

MARCH 4, 2021—CRIMINAL JUSTICE

> Andrea Roth, University of California, Berkeley, Law
> Renée Hutchins, University of the District of Columbia School of Law
> Sarah Brayne, The University of Texas at Austin
> Jens Ludwig, University of Chicago

MARCH 11, 2021—WORK AND LABOR, PART I

> Ece Kamar, Microsoft Research
> Min Kyung Lee, The University of Texas at Austin
> Karen Levy, Cornell University

MARCH 16, 2021—HEALTH CARE, PART I

> Madeleine Clare Elish, Google, Inc.
> Amy Fairchild, The Ohio State University
> Latanya Sweeney, Harvard University
> Robert Wachter, University of California, San Francisco

APRIL 29, 2021—WORK AND LABOR, PART II

Mary Kay Henry, Service Employees International Union

MAY 6, 2021—HEALTH CARE, PART II

Vickie Mays, University of California, Los Angeles
Susan Cochran, University of California, Los Angeles

MAY 11, 2021—CIVIL JUSTICE

Gillian Hadfield, University of Toronto
Ben Barton, University of Tennessee College of Law

MAY 11, 2021—HEALTH CARE, PART III

Kevin Fu, U.S. Food and Drug Administration
Maryann Abiodun (Abi) Pitts, Stanford University School of Medicine/Santa Clara Valley Medical Center

MAY 25, 2021—PUBLIC GOVERNANCE

Ryan Calo, University of Washington
Daniel Ho, Stanford University
Ben Green, University of Michigan
Eden Medina, Massachusetts Institute of Technology

JUNE 10, 2021—INDUSTRY RESEARCH, PART I

Eric Horvitz, Microsoft Research

JUNE 24, 2021 — INDUSTRY RESEARCH, PART II

Greg Corrado, Google, Inc.
Mitchell Baker, Mozilla Corporation

JUNE 29, 2021 — GOVERNMENT RESEARCH SPONSORS

Sara Kiesler, National Science Foundation
Nina Amla, National Science Foundation
Phillip Root, Defense Advanced Research Projects Agency
Michael Lauer, National Institutes of Health

C

Federal Computing Research Programs Related to Ethical and Societal Impact Concerns

The National Science Foundation's (NSF's) Computer and Information Science and Engineering (CISE) Directorate and the Networking and Information Technology Research and Development National Coordination Office (NITRD NCO) were asked to provide examples of federal computing research programs announced or continued in fiscal year 2021 that include a call for research aimed at identifying and confronting ethical and societal concerns related to the computing research being proposed and factors to be considered in addressing them, including multidisciplinary research aimed at identifying ethical and societal concerns related to computing research. Programs apparently focused only on research ethics and integrity or regulatory compliance were removed from the lists the agencies provided.

PROGRAMS IDENTIFIED BY THE NSF CISE DIRECTORATE

Programs Led by CISE

- Computer and Information Science and Engineering: Core Programs. Through its core programs, the NSF CISE Directorate supports research and education projects that develop new knowledge in all aspects of computing, communications, and information science and engineering, as well as advanced cyberinfrastructure. Issues of fairness, ethics, accountability, and transparency (FEAT) are important considerations for many core topics in computer and information science and engineering. In projects that generate artifacts ranging from analysis methods to algorithms to systems, or that perform studies involving human subjects, PIs are encouraged to consider the FEAT of the outputs or approaches. CISE is also interested in receiving proposals whose primary foci are on methods, techniques, tools, and evaluation practices as means to explore implications for FEAT. Among the CISE core programs, the Human-Centered Computing (HCC) program supports research in human–computer interaction (HCI), taken broadly, including the assessment of benefits, effects, and risks of computing systems.

- Cyber-Physical Systems (CPS). CPS are engineered systems that are built from, and depend upon, the seamless integration of computation and physical components. The CPS program aims to develop the core research needed to engineer these complex cyber-physical systems, some of which may also require dependable, high-confidence, or provable behaviors. Core research areas of the program include human-in- or human-on-the-loop, safety, security, and verification.

- Designing Accountable Software Systems (DASS). The DASS program solicits foundational research aimed toward a deeper understanding and formalization of the bi-directional relationship between software systems and the complex social and legal contexts within which software systems must be designed and operate.

- Expeditions in Computing (Expeditions). The Expeditions program is designed to inspire the CISE research and education community to be as creative and imaginative as possible in the design of bold projects that explore new scientific frontiers. Projects must describe policies on intellectual property and ethics.

- Formal Methods in the Field (FMitF). The FMitF program aims to bring together researchers in formal methods with researchers in other areas of computer and information science and engineering to jointly develop rigorous and reproducible methodologies for designing and implementing correct-by-construction systems and applications with provable guarantees. This includes verification techniques for machine-learning systems that could provide assurances of safety, correctness, and fairness.

- National Artificial Intelligence (AI) Research Institutes. The National AI Research Institutes program is a long-term multi-sector initiative to enhance innovation through foundational and use inspired research. Projects must include an ethics plan that provides a clear statement of the Institute's policies on ethics training, responsible conduct of research, and intellectual property rights.

- NSF Program on Fairness in Artificial Intelligence in Collaboration with Amazon (FAI). The FAI program supports computational research focused on fairness in AI, with the goal of contributing to trustworthy AI systems that are readily accepted and deployed to tackle grand challenges facing society.
- Secure and Trustworthy Cyberspace (SaTC). The SaTC program aims to protect and preserve the growing social and economic benefits of cyber systems while ensuring security and privacy.
- Smart and Connected Communities (S&CC). The S&CC program accelerates the creation of the scientific and engineering foundations that will enable smart and connected communities to bring about new levels of economic opportunity and growth, safety and security, health and wellness, accessibility and inclusivity, and overall quality of life. The program encourages researchers to work with community stakeholders to identify and define challenges they are facing, enabling those challenges to motivate use-inspired research questions.
- Smart Health and Biomedical Research in the Era of Artificial Intelligence and Advanced Data Science (SCH). The SCH program supports research with the promise of disruptive transformations in biomedical research drawing from multiple domains of computer and information science; engineering; mathematical sciences; and the biomedical, social, behavioral, and economic sciences. Program themes include information infrastructure; maintaining sensitivity to legal, financial, cultural, and ethical issues; effective usability, taking into account ethical, behavioral, and social considerations; and unpacking health disparities.

Programs Led by Other NSF Directorates

- Ethical and Responsible Research (ER2). The ER2 program funds research projects that identify (1) factors that are effective in the formation of ethical science, technology, engineering, and mathematics (STEM) researchers and (2) approaches to developing those factors in all STEM fields that NSF supports.
- Future of Work at the Human Technology Frontier (FW-HTF). The FW-HTF program supports multi-disciplinary research to sustain economic competitiveness, to promote worker well-being, lifelong and pervasive learning, and quality of life, and to illuminate the emerging social and economic context and drivers of innovations that are shaping the future of jobs and work.

Other NSF Activities

- Vulnerability Disclosure Policy. NSF welcomes the research and assessment of potential IT security vulnerabilities from independent researchers. The agency's Vulnerability Disclosure Policy offers guidelines for conducting vulnerability discovery activities about NSF and conveys agency preferences in how to submit discovered vulnerabilities to NSF.

PROGRAMS IDENTIFIED BY THE NITRD NCO

Department of Commerce

National Institute of Standards and Technology

- The National Institute of Standards and Technology (NIST) is developing a risk management framework to better manage risks to individuals, organizations, and society associated with AI. A July–September 2021 RFI and October 2021 workshop focused on trustworthy AI and on addressing technical and societal challenges and risks to AI.
- NIST AI Fundamental Research—Free of Bias project aims "to understand, examine, and mitigate bias in AI systems."
- NIST Text Retrieval Conference includes a track on Fair Ranking, which evaluates systems according to how well they fairly rank documents. The project's goal is to find "appropriate ways to measure the amount of bias in data and search techniques." Once determined, they will focus on identifying strategies for eliminating bias.

Department of Defense

Defense Advanced Research Projects Agency

- In Section 7.2.3, "Existing DARPA Efforts to Manage ELSI Concerns," the National Academies report *Emerging and Readily Available Technologies and National Security: A Framework for Addressing Ethical, Legal, and Societal Issues*[1] discusses the Defense Advanced Research Projects Agency's (DARPA's) understanding of the "issues surrounding the ethical, legal, and societal framework" of research.

---

[1] National Research Council and National Academy of Engineering, 2014, *Emerging and Readily Available Technologies and National Security: A Framework for Addressing Ethical, Legal, and Societal Issues*, Washington, DC, The National Academies Press, https://doi.org/10.17226/18512.

- As reported by program manager Brian Kettler in DARPA's Information Innovation Office, the Influence Campaign Awareness and Sensemaking (INCAS) program, which is designed for developing "techniques and tools that enable analysts to detect, characterize, and track geopolitical influence," can help researchers recognize when political influence affects results.
- DARPA's FY 2022 budget request includes the Computers and Humans Exploring Software Security program, to "enable computers and humans to reason collaboratively over software artifacts ... with the goal of finding vulnerabilities more rapidly and accurately than unaided human operators."

National Security Agency

- The National Security Agency's (NSA's) Civil Liberties, Privacy and Transparency (CLPT) director advises NSA senior leadership on the protection of civil liberties and privacy. The CLPT director is the lead for "promoting and integrating civil liberties and privacy protections into NSA policies, plans, procedures, technology, programs, and activities."

Office of the Secretary of Defense

- In a May 26, 2021, memo, Deputy Secretary of Defense Dr. Kathleen Hicks lays out the Department of Defense's (DoD's) AI Ethical Principles and establishes the implementation of Responsible AI (RAI) in the Department.
- DoD's Joint Artificial Intelligence Center (JAIC) has announced the Responsible AI Procurement "pilot of a procurement review process that will ensure AI acquired by the JAIC is aligned with DoD's [sic] AI Ethics Principles," as established by Hicks's memo regarding responsible AI.

Department of Energy

- In its Advanced Scientific Computing Research (ASCR) program narrative, the Department of Energy Office of Science (SC) stresses its commitment "to advancing a diverse, equitable, and inclusive research community." To that end, SC discusses the ESnet, which has moved toward shrinking the gender gap in scientific research. SC also states that ASCR will be participating in the RENEW initiative, which provides training opportunities to students in under-served communities.

Department of Health and Human Services

National Institutes of Health

The National Institutes of Health (NIH) Strategic Plan for Data Science aims "to balance the need for maximizing opportunities to advance biomedical research with responsible strategies for sustaining public trust, participant safety, and data security" and "[i]mprove the education of students on NIH training grants by enriching content in Responsible Conduct of Research requirements with information about secure and ethical data use."

The AIM-AHEAD Coordinating Center is designed to "increase the participation and representation of researchers and communities currently underrepresented in the development of AI/ML models."

The Office of Data Science Strategy announced that NIH is funding a new consortium to lead the AIM-AHEAD coordinating center, bringing "together experts in community engagement, AI/machine learning (ML), health equity research, data science training, and data infrastructure."

NIH issued an RFI on "current challenges and opportunities of using cloud computing at universities and colleges" to increase cloud computing access in "diverse biomedical research institutions." The RFI asks for responses to topics such as "barriers to adopt cloud computing including, but not limited to, training and infrastructure gaps, technical barriers, social challenges, perceived risks, and costs." It also requests information on "opportunities and potential impact on biomedical, clinical, behavioral and social science research from greater use of cloud computing."

Department of Homeland Security

Science and Technology Directorate

- According to Kathryn Coulter Mitchell, science advisor to the Homeland Security Secretary, the Department of Homeland Security is forming a partnership with the Cybersecurity and Infrastructure Security Agency, Critical Infrastructure Resilience Institute Center of Excellence, and industry innovators to provide retraining and reskilling in cyber security, as well as "drive new investments in diversity, social sciences, and research, development, and innovation, which are needed to build the next generation workforce."

Department of State

- The Department of State Office of S&T Cooperation facilitates American science collaboration and lays its ground rules, providing "valuable access for American scientists to foreign scientific capabilities, facilities, and expertise while also exposing other countries to American science procedures, norms, and values."